Get Your Amazing Invention on Store Shelves

An A-Z Guidebook for the Undiscovered Inventor

Michael J. Cavallaro

How to Get Your Amazing Invention on Store Shelves:
An A-Z Guidebook for the Undiscovered Inventor

Copyright © 2011 by Atlantic Publishing Group, Inc.
1405 SW 6th Ave. • Ocala, Florida 34471 • Phone: 800-814-1132 • Fax: 352-622-1875
Website: www.atlantic-pub.com • E-mail: sales@atlantic-pub.com
SAN Number: 268-1250

Library of Congress Cataloging-in-Publication Data

Cavallaro, Michael J.
  How to get your amazing invention on store shelves : an A-Z guidebook for the undiscovered inventor / by Michael J. Cavallaro.
     p. cm.
  Includes bibliographical references and index.
  ISBN-13: 978-1-60138-302-0 (alk. paper)
  ISBN-10: 1-60138-302-9 (alk. paper)
  1. Inventions--Marketing. 2. Patents. I. Title.
  T339.C38 2011
  658.8--dc23

                                    2011023634

Printed in the United States

PROJECT MANAGER: Gretchen Pressley • gpressley@atlantic-pub.com
BOOK PRODUCTION DESIGN: T.L. Price • design@tlpricefreelance.com
PROOFREADER: C&P Marse • bluemoon6749@bellsouth.net
FRONT COVER DESIGN: Shannon Preston
BACK COVER DESIGN: Jackie Miller • millerjackiej@gmail.com

A few years back we lost our beloved pet dog Bear, who was not only our best and dearest friend but also the "Vice President of Sunshine" here at Atlantic Publishing. He did not receive a salary but worked tirelessly 24 hours a day to please his parents.

Bear was a rescue dog who turned around and showered myself, my wife, Sherri, his grandparents Jean, Bob, and Nancy, and every person and animal he met (well, maybe not rabbits) with friendship and love. He made a lot of people smile every day.

We wanted you to know a portion of the profits of this book will be donated in Bear's memory to local animal shelters, parks, conservation organizations, and other individuals and nonprofit organizations in need of assistance.

— *Douglas & Sherri Brown*

PS: We have since adopted two more rescue dogs: first Scout, and the following year, Ginger. They were both mixed golden retrievers who needed a home.

Want to help animals and the world? Here are a dozen easy suggestions you and your family can implement today:

- *Adopt and rescue a pet from a local shelter.*
- *Support local and no-kill animal shelters.*
- *Plant a tree to honor someone you love.*
- *Be a developer — put up some birdhouses.*
- *Buy live, potted Christmas trees and replant them.*
- *Make sure you spend time with your animals each day.*
- *Save natural resources by recycling and buying recycled products.*
- *Drink tap water, or filter your own water at home.*
- *Whenever possible, limit your use of or do not use pesticides.*
- *If you eat seafood, make sustainable choices.*
- *Support your local farmers market.*
- *Get outside. Visit a park, volunteer, walk your dog, or ride your bike.*

Five years ago, Atlantic Publishing signed the Green Press Initiative. These guidelines promote environmentally friendly practices, such as using recycled stock and vegetable-based inks, avoiding waste, choosing energy-efficient resources, and promoting a no-pulping policy. We now use 100-percent recycled stock on all our books. The results: in one year, switching to post-consumer recycled stock saved 24 mature trees, 5,000 gallons of water, the equivalent of the total energy used for one home in a year, and the equivalent of the greenhouse gases from one car driven for a year.

Testimonial

Impressive in its sweep and precision, this book expertly explores the intersection of innovation and business. Now more than ever, today's inventors need to speak the language of Wall Street. The author's work does just that and provides a profound, coherent framework to increase chances for success. Along with biographies on Thomas Edison, inventors of all stripes should have *How to Get Your Amazing Invention on Store Shelves* on-hand at all times.

Mike Drummond
Editor
Inventors Digest
www.inventorsdigest.com

Author Dedication

*This book is dedicated to Hope
...and the possibilities she brings us.*

Table of Contents

Chapter 7: Understanding Patent Laws ..105

Chapter 10: How to Avoid Scams 183

Chapter 11: Coordinating Your Commercialization Strategy 197

Chapter 12: Foreign Markets 213

Introduction

On December 17, 1903, two inventors by the names of Wilbur and Orville Wright stood at the perch of the Kill Devil Hills in North Carolina with the intention of manning the first fully controllable aircraft in modern human history. From their vantage point, they could see a vacant shoreline patrolled only by birds that flew on the winds of the Atlantic. Sixty years after their first successful flight from those hills, December 17 would become a day commemorated by presidential proclamation to honor the two men whose invention altered the course of history forever.

The Wright Brothers were inventors, and like many inventors before or since, they were humbled by previous trials of miscalculation and experiments that ended in disaster. When it came to addressing the issue of aerodynamics, they knew they had to solve the problem of an aircraft's ability to stabilize itself in flight. No one before them had done it, so they had to study and re-examine their mistakes, find a new approach, and ultimately, *invent*. As a result of those labors, they designed a revolutionary device known as the three-axis control, which to this day remains the standard on all fixed-wing aircraft.

Innovation happens when the will to resist an obstacle finally outstrips the immovable walls of obstruction. The result is a tangible product, such as a composition, device, or process the world can use. Innovation takes creativity and drive, but if you are reading this book, you already know everything about that process and have either invented something or may be thinking about inventing something. In that case, your next question should be: How do I bring my invention to market? The answer to your questions lies within the pages of this book.

According to the U.S. Patent and Trademark Office, roughly 450,000 patent applications are filed for new inventions annually. Although this may seem like a lot, the number of successfully issued patents total far less due to the number of rejected applications that either fail to meet proper qualifications, violate some enforceable standard, or infringe on a similar patent.

Over the course of this reading, we will discuss how market research can help convert your idea or invention into a product that will be manufactured and funneled into a distribution chain that reaches customers. We also will cover how to protect you and your product from the financial and legal pitfalls associated with the process of commercialization. Finally, this book will examine how to profit from the deals you make with potential licensees who want to manufacture and distribute your invention, as well as how to profit by creating an independent enterprise.

The market research you conduct, and the people you encounter should provide a reasonable barometer as to where your invention stands as a viable commercial product. Simply put, be willing to ask yourself if the expectations you have for your invention hold up with the realities of the marketplace. If so, there is much to gain. Let's get started.

Owning Your Idea

If you have never thought of an idea as a form of property, start thinking that right now. When you develop ideas into an original conception that has the potential to be bought and sold in the marketplace, that means you are the original owner of that conception and thus, own the intellectual property. The procedure that starts with the genesis of the idea, results in mass production, and ends with the customer buying the product is a process known as **commercialization**. Commercialization can be achieved by obtaining licensing deals with companies that acquire some aspects of your intellectual property for the right to make and sell your idea to the customer, otherwise known as the end user.

Universal First Steps

Before any type of commercialization, one must measure the uniqueness of the idea, its market potential, and the financial undertakings required. If all indications point toward an idea that can be realistically converted into a product, a governmental regulating body that specializes in intellectual property rights must grant legal protection of the intellectual property. In the United States, that regulating body is the U.S. Patent and Trademark Office (USPTO). Filing an application with the USPTO means you are asking them to acknowledge and grant your right to own the property.

This official grant is called a **patent**, which gives the inventor the right to prevent others from making, selling, or even using a product based on the claims made in the invention's patent. In the United States, a patent's protection lasts for 14 or 20 years. Patents are personal property that can be sold or licensed by the inventor.

Once you have sought patent protection, the next step is to acquire a manufacturing deal with a company willing to produce your product in volume, followed by a distribution deal that physically delivers the volume of that product to the marketplace. Forging deals with manufacturing and distribution companies most commonly involves licensing or selling them all or certain intellectual property rights to your invention. If they acquire some aspects of your property rights, they share your patent, and thus, some of the profits that will come with owning and selling the invention.

Inventors are creative, imaginative problem solvers. Often, they are technically knowledgeable in their fields but not savvy and sophisticated business people. The aim of this book is to educate the inventor to the ways of the business world so he or she can maximize the rewards of his or her creations.

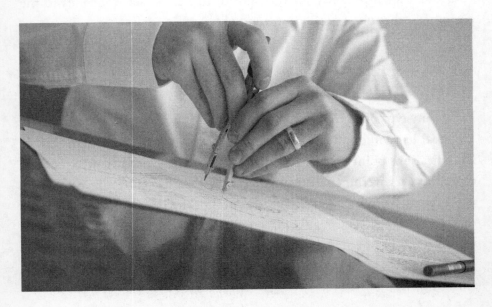

Some Things to Avoid

Some factors will knock your invention out of consideration. Make sure there are no disqualifying factors. If there are, save yourself the time and effort of marketing your invention. An invention is commonly disqualified from consideration because:

- It is illegal or violates industry standards or regulations.

- It moves the company, brand, or product line in a different direction than its current strategy.

- It requires unproven or expensive new equipment or technology.

- The initial investment outweighs the likelihood of profiting from the invention.

- The industry is leapfrogging over your invention's innovation, and it will not increase their competitiveness enough to justify commercialization.

Commercialization Strategy

Commercialization enables the inventor to get his or her invention into the hands of consumers. If your product is going to be available, it must first be made into a product that can be manufactured, distributed, and sold to consumers. A sound commercialization strategy will optimize the chance your invention will be developed successfully into a product that ends up on store shelves and in consumers' hands. Developing a commercialization strategy is a multilayered process that involves research, industry feedback, monetary investment, and time. Once you implement your strategy, you may find there are areas where more detailed research is necessary, and you will likely follow up again to further refine some critical areas of your strategy.

It is hard to know where to begin with an invention. There is no universal approach right for every concept. Some ideas should be patented right away, others patented only when a partner is interested in investing in or licensing or purchasing the idea, and in some cases, you will not need a patent at all. Some inventions will be sold or licensed based on a model or prototype. Others will gain positive attention of investors or potential licensees after you have created the product yourself and test-marketed it.

Steps to commercialization

Although property protection strategies among inventors may differ depending on the nature of their product, a natural progression in the process of commercialization is universal to every invention. When you develop your commercialization strategy, adhere to the following ten-step process:

Step 1: Research your product and industry.

Step 2: Identify and research your consumers, market, and industry.

Step 3: Develop a preliminary strategy outline.

Step 4: Document your invention.

Step 5: Evaluate your invention's commercialization potential.

Step 6: Avoid invention-marketing scams.

Step 7: Select a commercialization route.

Step 8: Identify intellectual property protections likely to cover your invention.

Step 9: Choose a manufacturing and marketing route.

Step 10: Find appropriate partners and execute necessary agreements.

Preliminary strategy outline

Before you decide to apply for a patent, you have to know which intellectual property rights to protect. You will come to know which rights to protect by having a commercialization strategy in mind. A **commercialization strategy** is defined as a series of business decisions an inventor must make to move an invention from concept to marketplace. This book has been set up to provide a chronological overview as to how that process unfolds; therefore, the knowledge you gain should be the foundation of your strategy. Start by defining what your product does and what industry or market segmentation it appeals to. Once you have identified the industry, begin with an initial screening by consulting with experts in that industry, such as industrial designers who will be able to provide a preliminary marketing and technical evaluation assessment of your invention. *Industrial designers will be discussed in Chapter 3.* Industry consultants also will be able to define who the major and minor players in that industry are and what the estimated cash requirements for patent applications, marketing, and legal services will be. When this has been established, set general timelines to meet each of your goals. Commercialization strategies also can be represented by reaching all the milestones of the entire commercialization process, which include obtaining a preliminary market assessment, creating a product prototype, connecting with manufacturers interesting in market testing, achieving property protection, obtaining license deals, and moving into other markets.

The importance of a commercialization strategy is twofold. First, you want to have a realistic plan that concretely and systematically sets reachable goals that deter operational pitfalls. The second reason is financing. If you do not have the proper cash requirements, you will have to find a lender who believes in the viability of your invention. Lenders will refuse to lend money out of blind faith. They want to know that your product stands a realistic chance of being commercialized, and the money they are lending has a high probability of being paid back. The more focused the strategy, the greater the chance of convincing them the principle amount will be paid back at the interest rate you agreed to pay.

Gathering strategic information on your invention's industry is relevant and will assist you in making several key decisions. Examine the products within your industry. Conduct market research to find out how similar products were manufactured, how they were distributed and marketed, and what their sales numbers look like. How large is the market? How many competitors are in the market? What are their individual market shares?

Industry research is the best way to determine if your idea is patentable. Inventions must be useful, new, and nonobvious to be considered patentable by the Patent and Trademark Office's examiners.

> According to the USPTO, to be patentable, an idea or invention must "be nonobvious to a person having ordinary skill in the area of technology related to the invention." In other words, the average person who is aware of all the historical and current work in your field should not be automatically able to think of your invention idea. Ask yourself: is your invention useful, new, and nonobvious? If so, your invention may be patentable.

To receive a patent, you will have to prove in your application that your invention meets these criteria. If you have one or more applications in mind for your invention, you will be able to establish utility. There does not have to be a product available in the stores based on this patent for your invention to be unpatentable. Patents are reserved for new and unique products, devices, formulas, or processes. If someone already has expressed the idea of your invention or holds a patent on the idea you have conceived, it is not novel, and you cannot patent it. To gauge the novelty of your idea, conduct your own preliminary patent search, otherwise known as **prior art search**. There are several ways a prior art search can be conducted.

The first way is to hire a patent attorney to conduct the search. Patent attorneys usually charge fees between $500 and $1,200. If the price range feels too steep, a less expensive way to search prior art is to hire a *patent agent*. The difference between a patent attorney and a patent agent is the level of professional experience. A patent agent is someone who has passed the bar exam and is registered with the USPTO, whereas a patent attorney has obtained a graduate degree in law called a Juris Doctor, has passed the bar and patent bar exam, and is registered with the USPTO. A third way to conduct a prior art search is to do it yourself. This can be done through online databases such as:

- Dialog®: **www.dialog.com**
- Free Patents Online: **www.freepatentsonline.com**
- Google Patents: **www.googlepatents.com/patents**
- Delphion: **www.delphion.com**
- USPTO Patent Database: **http://patft.uspto.gov**

When a patent is issued, the information you submitted in your patent application becomes public. A patent can give competing entities critical information about your idea that may embolden them to create a similar product. If your invention obtains certain patent rights competitors can easily "engineer around" without penalty, a patent may have little or no use to you. This is what makes knowing which rights to obtain in a patent so important. A patent is made up of one or more claims depending on the complexity of your invention, and each claim is made up of one or more elements. A patent's design claims protect you from anyone else using all of these claims and elements together. If someone can streamline your process and omit a single element, he or she can circumvent your patent and render its protection useless. A patent with superfluous claims or elements is weak and can easily be circumvented. The best way to create a strong claim is use prior art search to determine what claims are already taken, which claims are similar, what sets your claims apart from existing claims, and which forms of protection will broaden the scope of your property rights.

Patents are expensive to obtain. Nonprovisional or regular utility patents average around $12,000 each in legal and filing fees. If your invention is not certain to earn you more than this, it makes more sense to seek a **provisional patent**, which is less expensive and can protect your idea while you look for a licensee. If you license your invention with a non-patent form of protection, your licensee can always pay to pursue a patent in your name if they believe patent protection will enhance their competitiveness. Even if you are convinced a patent will protect your property, do not fill out the application until you have determined how a patent fits in your overall commercialization strategy. Examining your personal strengths and weaknesses will help you determine what type of outside help or expertise to seek during commercialization. Ask yourself the following questions:

- Are you committed and feel competent enough to research all the legal aspects of intellectual property rights and licensing agreements?

- Do you want direct involvement in manufacturing, distribution, and marketing?

- Are you an expert in the industry your product is attempting to penetrate?

- Do you have any experience negotiating deals?

- Do you have any marketing experience?

If you do not intend to assign rights over to anyone permanently, you will have to seek investors and other partners with manufacturing, design, and distribution capabilities. Some inventors are technical experts in a specialized field and have access to the materials, equipment, and expertise necessary to create a professional design and produce a model or prototype; others are laypersons with innovative ideas. For laypersons, designing prototypes may require the knowledge of a professional, such as an industrial designer.

Documenting Your Invention

For reasons pertaining to intellectual property rights, documenting your invention should start as soon as you have come up with something to invent. Doing so proves you have conceived the idea at the earliest possible date. Documentation supports any future claims you may have made to protect your invention. You also may need documentation in the event of infringement on a trade secret if you decide not to patent and someone else improperly makes your secret public. Buyer beware: new legislation to reform patent laws through the America Invents Act as discussed in Chapter 12 may soon reduce the benefits of documentation with respect to foreign patent applications. It should, however, be noted that documentation of an invention is still crucial for design and prototype purposes.

Documentation should come in the form of keeping notes on any ideas, tests, and developments involving your invention. It is also used to support development of complex ideas by tracking avenues of exploration and their results to prevent repetition and backtracking. More important, the IRS may want to know you are working on an invention when you submit claims for business expenses that have nothing to do with your day job. Using a notebook to track the process will suffice and may be any bound volume of blank, lined, or graph paper. The purpose of the notebook is to establish that you were the originator of the idea, and its conception occurred to you before another person who may try to claim him or herself as the original inventor in a court of law.

How to document

To document your invention, write a detailed description of your invention as soon as the idea occurs to you. If relevant, documentation can include a descriptive narrative, calculations, equations, formulae, sketches, diagrams, or visual depictions. Take each piece of documentation and chronologically number, sign, and date each entry of the write-up. If you use a bound notebook, keep it in a safe place where no one can access it. Documentation may become useless if you disclose your idea to anyone more than one year before patent application, and your invention becomes ineligible for a patent.

If one entry is several pages long, sign and date each page. A bound notebook enables you to document a chain of events and ideas that occurred in a plausible chronological sequence. Loose-leaf material does not permit building the same sort of case. Each time you refine through experimentation, you must document that event in a notebook. Once the idea has been fully developed, draft a confidential disclosure agreement and a witness statement. The witness statement further validates your documentation and the invention process. Your documentation should be witnessed by an impartial but trusted outside party as confirmation of your work. Draft and date a witness form with a detailed description of your invention, experiment, or process involving your invention. Using this witness form, have your witness sign under a written statement saying he or she understands what he or she has witnessed. These witnesses also should be made to sign a nondisclosure agreement to preserve your trade secrets as proof that the USPTO should not consider it public information, which would disqualify it for a patent. *There is a sample witness form and nondisclosure agreement in the Appendix.*

Your notebook should include all sorts of information about your invention. Begin with a description of the idea, how you conceived it, and how you envision it working. If you are not confident in your ability to draft sketches of the invention, contact a professional sketch artist and ask them to sign a nondisclosure agreement. *See Chapter 5 and Appendix C for further details.* If you discuss your idea with witnesses, do not disclose any more information than necessary to create a witness form. Keep documenting as you refine your invention through research and experiments. If you are building your own prototype, document this activity as well.

CASE STUDY:
THE VALUED IDEA

Lisa Lloyd, president and business
strategist
Lloyd Marketing Group, Inc.
1 Innovation Way, Newark, DE 19711
lisalloyd@lloydmarketinggroup.com
www.lloydmarketinggroup.com
Phone: (520) 722-9545
Fax: (520) 722-2840

Since I became an inventor, I have licensed seven different products to companies in the grooming industry. I also consult for major corporations that offer licensing deals to develop new businesses with intellectual property. I lecture at business schools, was once a guest on *Dr. Phil*, and have been featured in magazines such as *Women's World*. I am currently the spokesperson for IdeaTango.com, which is an inventors' resource website with blogs and discussion forums. Right now, I am working on the launch of a new product line called Treasure Chest Pets.

So, how did I come to have such a full schedule? It started when I licensed my first invention, the French Twister™, to a major hair products manufacturer. Like many inventors' ideas, the idea came from a problem I had trouble solving. I often wore my hair in a French twist to keep it neatly groomed for work. As my hair grew longer, it became more difficult to keep it styled upward. After a fruitless search for the right type of barrette, I decided to invent one myself.

I did a lot of research to determine the scope of each industry, what was already being done, the cost to produce goods, and an estimation of sales based on similar products. My mother was a prominent business-woman in her community with good contacts, so I leaned heavily on her for advice in the early going.

I used a very successful hair clip inventor by the name of Tomima Edmark as a business model I would try to emulate and then created a three-step process for commercializing the French Twister. First, I scoured the stores for lines that my product would make sense in. Second, I contacted the manufacturers and distributors listed on the back of the packaging. Third, I got their submission procedure and developed a prototype.

I wrote my first business plan at the Tucson Small Business Development Center, a local office of the U.S. Small Business Administration. After that, I applied for design and utility patents, spending about $50,000 to commercialize the French Twister but almost nothing on subsequent hair accessory inventions because after that there were no learning curve expenses. Ultimately, I earned $500,000 on the French Twister.

I was 23 years old at the time, so I really wanted to build my own business, but licensing had such a great return on investment that it did not make sense to start a one. Now that I have my own business, I have recently expanded from hair accessories and have begun licensing a patent from another inventor for a line I intend on selling to stores.

Over the years, I have become more of a researcher. I am not willing to lose time or money on ideas that are not profitable enough. The proper amount of research is the only way I believe you can avoid costly mistakes. I primarily use the Internet, but I am also an avid reader of *Entrepreneur*, *Fortune*, and *Fortune Small Business*. I believe it is equally as important to stay encouraged.

Evaluating Your Invention

Although all inventions take creativity and drive, not all inventions will have a place in the commercial market. As the inventor, you need to find out if it is worthwhile to put more money into a certain idea. The only objective way to determine an invention's commercial potential, functionality, and design is to get a professional evaluation. Receiving an evaluation is especially important when trying to license ideas or raise venture capital because receiving an honest and expert evaluation will later become a key cog in your pitch to these potential suitors. Before seeking an evaluator, ask yourself certain questions regarding your product, such as:

- Is there a need for your invention?
- Does your invention work?
- Is it safe, legal, and ethical?
- Do the projected manufacturing and distribution costs still allow you to price the product competitively?
- Is it an appealing product and a good value for the consumer?
- How much engineering and design work remains before your invention can be turned into a product?
- How much capital is needed to get the invention produced?
- How many people or companies will want to buy the product?

- Is demand for products in this industry or niche growing or waning?
- How many other products are on the market meeting the same need?
- Is your product easy to use, and is it obvious how it works?
- Is this product durable?
- Once your product is introduced, will others saturate the market with similar products?
- How novel is your idea? Will consumers embrace it or have to be sold on it?
- Will your product stand out in a crowded marketplace and grab consumer attention?
- Does your product have the potential to be turned into a line of products?
- How much more research and testing are required before your invention is ready for the market?

It is also important to consider your own ambitions. Ask yourself the following questions:

- Is recognition driving your ambition, or is money?
- Are you hoping to make your first million from this invention, or are you interested in learning about the process to develop future products?
- Are you planning to work at this full time?
- Are you willing to invest money in this invention?
- How many years are you willing to work at getting this invention into consumers' hands?

Once you have defined your own parameters, you will be more equipped to determine whether your commitment is worthwhile for this particular invention.

The Importance of Early Evaluation

The invention you create may be entirely your own, but when it comes to commercializing your invention into a marketable product, you will have to trust the experts you hire to help bring your product to market. Before you spend any time, effort, or money pursuing commercializing an invention, you need to have experts critically evaluate your invention. Many inventors in the past have made the mistake of showing their prototypes to potential licensees before they have assessed their products' value to customers in the marketplace.

When a manufacturer is considering the possibility of buying intellectual property rights, the first thing they will do is have their marketing department run a market research analysis to determine the product's potential market value and/or have its engineering department run tests on your prototype to determine its functionality. If their findings are of low value to the company, your offer will be rejected. To avoid this, consider having an evaluation of your invention conducted by experts who can provide their own analysis to identify any marketing, design, or functionality weaknesses. Their findings may identify certain weaknesses that may indicate your product will fail, in which case you may need to think about abandoning further pursuit of commercialization before you incur any more costs. However, if you have already conducted a preliminary commercialization strategy, you have good reason to believe that market potential exists. Therefore, most qualified experts will attempt to identify any problems facing your invention and offer solutions in order to give it the best chance possible of getting to the market and thriving there.

Before pitching an invention to a manufacturer, and perhaps before you incur any costs involved in pursuing a patent or building a prototype, consider getting an evaluation of your idea from a reputable evaluation service.

Invention Evaluation Services

In addition to private organizations, many universities, nonprofits, and government organizations also offer evaluations. Keep in mind that these organizations are more likely to be legitimate than any for-profit organization. In fact, some of these organizations will evaluate your invention idea free and then elect to offer ongoing technical assistance for a flat fee or hourly rate. Other universities, nonprofits, and government agencies may offer this type of service for a few hundred dollars.

It may also be worthwhile to ask public and private organizations for the names of reputable invention agents who can facilitate the patent application process when the time to file becomes appropriate. Although agents typically focus on patent applications and work for a percentage of your licensing earnings, some agents will provide fee-based invention marketability assessments for roughly $200 in addition to their regular services. These organizations are good resources for independent inventors because they regularly file patent applications. Universities, for example, earn substantial royalties from employees who work inside their research and development programs, many of which are funded by government and nonprofit organizations. Some of the more notable university, nonprofit, and government invention services include the following organizations:

- The Washington State University Innovation Assessment Center offers invention evaluation and, for selected inventions, design and marketing assistance. It can be reached at **www.business.wsu.edu/organizations/iac/pages/index.aspx**.

- The Wisconsin Innovation Center at the University of Wisconsin at Whitewater offers invention evaluation. It can be reached at **http://wisc.uww.edu**.

- The United Inventors Association Innovation Assessment Program provides invention evaluation using a structured evaluation method for $300. It can be reached at **http://uiausa.org**.

- The Baugh Center for Entrepreneurship at Baylor University offers an innovation evaluation service for a fee that covers its expenses. It can be reached at **www.baylor.edu/business/entrepreneur**.

- The Bucknell University's Small Business Development Center assists Pennsylvania-based inventors in creating new products and exporting products, provides patent assistance, and helps inventors find markets for their products. It can be reached at **www.bucknell.edu/x7025.xml**.

- The Maine Patent Program at the University of Maine School of Law provides invention evaluation and assistance for Maine residents only. It can be reached at **http://mainepatent.org**.

- Public Interest Intellectual Property Advisors offers advice for inventors seeking to promote projects in the developing world. They can be reached at **www.piipa.org**.

- The National Institute of Justice's Office of Law Enforcement Technology Commercialization in Wheeling, West Virginia, provides assistance for inventors of technology with law enforcement or corrections applications. It can be reached at **www.wvhtf.org/departments/public_safety/programs/oletc.asp**.

- The Canadian Innovation Centre provides assistance with Canadian patents. It can be reached at **www.innovationcentre.ca**.

Vet these organizations carefully, as nothing guarantees their legitimacy or success if you use them. The best approach is to ask for a list of clients and clients' individual sales numbers, that is, how many inventions they have sold. If they refuse to provide information, contact someone else. As you work your way down the list of evaluators, keep in mind the sobering fact that statistics do not favor your invention. Although patents have been issued for seemingly every imaginable and unimaginable device, few have gone on to become profit-making products. According to the U.S. Patent and Trademark Office, only 2 percent of patents earn money for their inventors. Even if the invention is guaranteed to provide consumer value, whether by saving money or enriching lives, profits are determined by external factors that put most inventions at the mercy of the marketplace's whims. If you decide to evaluate the invention yourself, online resources that provide evaluation instruction and checklists can be found at:

- Virtual Pet: **www.virtualpet.com/invention/inventionprocess.htm**

- Entrepreneur: **www.entrepreneur.com/encyclopedia/checklists/article81922.html**

Evaluating Your Invention

According to the Innovation Institute (**www.wini2.com**) there are 45 specific criteria, subheaded under six major categories, that determine the benefits and risks involved in commercialization of potential products. Its evaluations are designed to aid the inventor's decisions and include strategies for further development. A critical evaluation costs only $200 and may save you the headache of spending much more for patent applications and for a prototype that fails to become commercialized for all the wrong reasons. The six major categories of its evaluation are:

- Social criteria (your product's societal impact)
- Business risk (your product's financial benefit to licensees)
- Demand analysis (the current and projected environment of your product's market)
- Market acceptance (your product's likelihood of being accepted by customers)
- Competitive criteria (the level of competition your product will face)
- Experience & strategy criteria (recommendations on how to proceed in successfully commercializing your product)

When an evaluation service is conducting an analysis of your invention, the evaluator is going to test for a number of market factors. The factors taken into account are the aspects that manufacturers will consider when they evaluate your invention. If they pass the litmus test, they will offer to license your product for the right to make and possibly distribute it. Under social criteria, the first factor to be measured is the product's legality. A reputable service wants to make sure you are not going to be tied up in court because someone has contested your legal right to make and distribute your invention. An evaluator's report should include research on all applicable laws pertaining to liability, state and federal regulations, and product standards.

The next examination under social criteria is the product's safety. Does manufacturing your product require using hazardous materials? Is the use of these materials regulated or does it in some way violate environmental protection standards? If so, a detailed analysis of state and federal standards to be obeyed should be included in the report. If your invention violates these standards, the firm will work to help you find another solution to manufacturing the product that adheres to the law.

When considering business risk, evaluators are going to test the feasibility of the invention's functionality, which may include technical processes required for production. Determining functionality gives the evaluator a pathway into estimating the type of research and development that will be required to reach the production stage. From there, they can begin to estimate production costs, and by determining production costs, they can begin to compile a market research report. The market research report will cover all demand analysis criteria and should provide insight into your invention's potential market, projected sales, level of market demand, and life cycle.

Compiling an analysis of your invention's potential demand allows the evaluator to determine the critical factors that involve gauging your customer's willingness to accept your invention as a product. Customer acceptance relates to factors that influence buying decisions, such as a product's compatibility with the customer's existing usage of things, how long it takes the customer to learn its usage, and how necessary your product's integration has become with products the customer already depends on.

If your invention is not patentable, it is possible similar products have been created and competition already exists. In such cases, evaluators will examine competitive criteria, such as the design of your product as compared to the design of existing similar products. They may compare its durability, suggested retail price, and patent protection, if applicable.

Market appeal

The design of your prototype is a feature that professional evaluators examine because customers will use it as a deciding factor when weighing their options in the marketplace. If your product looks strange or clunky, you have a major barrier to overcome. If it is sleek, sexy, refined, or visually pleasing, its design will establish a competitive advantage. Industries that command a large market tend to favor products that are marginally better than their competition. Sometimes marginally better means nothing more than more attractive product design. This means the product commands higher market appeal.

To increase market appeal, your product should be more reliable, durable, easier to use, safer, quieter, less polluting (if regulated by EPA), more portable, better looking, pleasing to the touch, faster, or more efficient. Prior art searches should have information included if there is an EPA issue on similar claims. Also, a patent examiner hired by the inventor would check to make certain there are no regulations involved. *Prior art research is covered later in Chapter 6.*

If a product cannot claim advantages over the competition, it does not have a competitive advantage. If you are going to succeed as an inventor, you must focus on the marketing advantages of your invention after the evaluation process is complete. How will it sell itself? Will people be motivated to buy it? If you want to reach consumers, you have to focus on features that increase market appeal.

Inventing a marketable product is the key to successful commercialization. You can only profit from your idea if you convince a prospective licensee that they will make money. Some industries where there is little distinction between competing brands (laundry detergents, beverages, clothing lines) are moved primarily through marketing. These industries distinguish their products by using certain aspects of market appeal.

At the end of the analysis, your evaluator will work up a strategy based on the findings and make suggestions as to what your strategy should be to optimize your product's ability to reach the marketplace and profit.

After the Critique

Following an evaluation, many inventors are afraid to absorb the realities facing their inventions because they do not want to be dissuaded from pursuing it. Only you can talk yourself out of pursuing your dream. Remember that the role of the invention consultant is to provide a realistic outlook of your invention and the industry climate facing your invention. The suggestions they make for tailoring your invention and your approach for success are designed to improve your chances, save you money, and optimize your ability to succeed. Should you choose to ignore the findings in their report, you may have tough sledding ahead.

Decision point

After you have developed a preliminary strategy outline by yourself or with the assistance of a professional evaluation service, it is time to ask if your invention is worth pursuing. Do you truly want to spend most of your free time and some of your income for the next couple of years working to commercialize and protect this invention? Does it have the potential to meet the goals you have set for it? If the answer is yes, proceed forward with your preliminary strategy, and begin contacting industry members to compile market research that is more comprehensive. *This will be discussed in Chapter 4.*

If the answer is no, consider the reasons. Is the timing not right? Is it ever going to be worthwhile to pursue this invention? Under what conditions will it become feasible? By setting this invention aside, you are not betraying yourself or your creative muse. You are operating within the constraints of the market. Your invention can wait until the conditions are right for it to be successfully commercialized and protected. In the meantime, keep a notebook and pen on your nightstand, in your glove compartment, near the shower, by the microwave, and in your briefcase. The next idea will strike; be prepared to capture it, so you can evaluate it as well.

Competitive Advantage

Your product does not need a niche to possess some type of market value. Nevertheless, it must be better, faster, or cheaper than similar products on the market. When your invention can produce the same results cheaper, better, or faster than competition, that means you have a competitive advantage. Having a competitive advantage increases the chances of exceeding average industry profits and capturing larger market shares over your competitors. In other words, competitive advantage equals success.

Another factor that will measure your product's competitive advantage is how motivated people will be to change their buying habits because of your invention compared to the current consumer trends. When innovations occur, customers typically resist change until they are forced to adjust their consumer habits to the overall changes that certain products bring to the marketplace. If external pressures within the macro world are great enough to persuade them to change their buying habits, your competitive advantage increases. Gaining a competitive advantage is also about how your product's features can be marketed. Features that influence a product's market appeal may include niche, customer access, and price.

Products are thought to be more accessible when innovative technologies are added to an already existing platform. For example, when Microsoft® unveiled its search engine Bing®, it not only had more advanced features than its main competitor, Google, it was made easily accessible to consumers through Internet Explorer®, their existing platform of technology. Without an available browser and other widely used technologies, Bing would be considered an inaccessible product. When considering new products, evaluators will try to determine whether a product can maximize either profit margin or market share. An evaluator will measure the product against two pricing strategies: setting a high price and selling to customers who are less price sensitive, or setting a low price to achieve high sales volume. Evaluators believe that setting a high price is most appropriate when customers are believed to be less price sensitive, large cost savings are not expected at high volumes, or the manufacturer does not have the resources for high-volume production. Evaluators believe that setting a low price is optimal when customers typically choose lower-priced items, the cost to manufacture decreases in higher volumes, or competition is threatening a significant portion of market share.

When a product has found a specific, unique, and largely untapped part of an industry to target, we call that a **niche**. So, which individuals and companies should you target to achieve your niche, and which segment is going to be most interested? Do they have the money to buy or support your invention? If so, you may have found a worthwhile niche that few, if any, competitors are reaching. Determining market niche will play a role in the market research you collect following an evaluation. *This will be discussed in Chapter 4.*

CASE STUDY:
INDUSTRIAL DESIGNING

James E. Richardson, president
Richardson & Associates, Inc.
136 Goodwin Rd., Eliot, ME 03903
jamesranda@comcast.net
www.richardson-assoc.com
Phone: (207) 439-6546

I am an industrial designer who works with inventors to design and develop consumer product inventions. More specifically, I specialize in designing three-dimensional models and "proof of function" prototypes. Using my expertise, I help inventors find parts and materials for their models, as well as manufacturing and assembly partners. I also help them develop catalog-marketing sheets based on the models we create. When it comes to plastic injection molded products, inventors should not immediately decide the manufacturing technique. Tooling, design, and engineering are different processes, and inventors should trust the designer's suggestions. Commercialization is something you can learn; engineering is not.

I always recommend working with industrial engineers and designers who are familiar with the cost constraints inventors and small businesses face. Many engineers have experience doing projects in one industry where the cost is minimal, but it is a mistake to assume all industries are the same. If you have done your research, you know the product needs certain features and has to retail at a certain price. Using this information in conjunction with the amount you are willing to invest, an industrial designer can extrapolate your manufacturing options. The designer is coming up with a solution that fits the marketplace and your financial resources.

When I work for a client, I begin by asking the inventor for enough information to obtain preliminary price information from manufacturers. Then I will come back to inventors with information about what sort of tooling investment is required to make certain parts. If the price makes sense to the inventor, I have them collect information for a marketing plan, as well as a profile, that includes the product's features and benefits. The inventor also will have to research distribution channels, trade show dates, and catalog submission deadlines. Based on that information, I will write a work plan that lays out additional questions that need to be answered, including the subsequent phases of product development, such as drawing and prototyping.

If you have an injection-modeled part that you can have manufactured for $1 using a $40,000 tool, in some instances, you can get a $4,000 tool that will produce the product at $3 each. Even if the inventor has to manufacture and sell a run of the product at a loss, this sort of test run enables inventors to "test the market to raise certainty" before either investing in more expensive tools or using test data to market the product to licensees.

Building a Prototype

Building a prototype satisfies three purposes when attempting to bring your invention to store shelves. First, it will be used to help work out any possible kinks your invention may have once made into an actual product. Second, potential customers will use it to solicit interest when you begin to connect with key decision makers in the distribution tier. *See Chapter 4 for more information.* Third, it demonstrates your invention's **proof-of-function**. In other words, the prototype proves to the potential customer that your invention actually works when made into a product. If necessary, the prototype may have to undergo a series of redesigns based on the recommendations of the industrial designer you are working with. The more unique and complex an invention's proof-of-function is, the more likely the first prototypes will have to be redesigned to achieve a level that satisfies the potential customer. To build a prototype, you will have to hire a manufacturer or convince a potential licensee to build a certain amount of prototypes based on design and redesign.

Designing for the Manufacturers

Inventors always should be aware of the latest trends gripping the manufacturing sector. In recent years, manufacturers have started to hire companies that specialize in industrial design called **contract manufacturers**. These companies specialize in industries such as electronics, clothing, drug, plastics, and customization. The value of a contract manufacturer to a manufacturing company is its ability to manufacture or purchase all the components necessary for production. Some products, such as automobiles, are manufactured regionally in the places where they will be sold to reduce risks involved with currency exchange and transportation costs. In such cases, a manufacturer will outsource production to a contract manufacturer.

Contract manufacturers are good places to seek out tips on designing a prototype for a manufacturer because they already know what the manufacturer's design function requirements are. When your product is in the design or redesign stage, consider what type of design potential licensees may be looking for. As a frame of reference, it is useful to know the basic sequence of steps firms use to conceive and design a marketable product. During the product design process, a manufacturer will perform eight steps:

Step 1: Consider existing platform and assess possibility of new integration

Step 2: Assess feasibility of concepts and build a prototype

Step 3: Generate alternative architectures and/or refine prototype

Step 4: Select the best materials among original and alternative architectures, complete design, and retain documentation

Step 5: Conduct a performance and durability test

Step 6: Obtain regulatory approval

Step 7: Implement redesign to meet regulatory approval, if necessary

Step 8: Evaluate performance of final product design

Industrial Designers

Although some inventors have enough design expertise and resources to build their own prototype, others may possess neither. For inventors with skills limited to conceptualization of an idea, the first step in creating a prototype is to hire an industrial designer. Industrial designers are professional engineers who work with inventors and manufacturing firms to create products that are stylish, functional, and cost-effective to manufacture. They keep in mind the consumer's interests and needs, as well as the manufacturer's interests and needs. Industrial designers begin with product's parameters and constraints, which include funding and equipment capabilities, and proceed to design accordingly. Although the industrial designer can oversee and consult with you on the design and redesign of the prototype, the actual making of the prototype will be farmed out to the manufacturing company of your choosing. An industrial designer's priority includes looking for functionality, aesthetic appeal, safety, and manufacturing feasibility. Because industrial designers most often work for manufacturers, they may not be geared toward the unique needs of independent inventors. The Industrial Design Society of America's website has a directory of industrial design firms that can be searched by industry at **www.idsa.org/expert-design-witness-directory**.

When contacting design firms, be sure to ask whether they have worked with independent inventors before and request references. Simpler products such as zip-close bags involve less complex designs and are easier and cheaper to manufacture. However, more complex designs such as automotive navigation devices require engineering expertise for a finished product that meets inventor and manufacturer requirements.

Your Prototype

The prototype is a proof of function; it allows the inventor to demonstrate design and functionality features that are difficult to demonstrate in two dimensions. Inventors know what the product will look like, what parts will be needed, and how different elements will be integrated to achieve functionality. However, building a prototype often proves the inventor's initial design schema wrong. Although prototypes give the inventor a tangible proof of function to shop around, the more important purpose of building prototypes is to facilitate the process of solving design and functional issues that typically impede the original conception from becoming a viable product. Furthermore, prototypes help the inventor measure the initial level of customer interest. The true value of a prototype, however, lies in the more critical aspects of the commercialization strategy — the phase in a preliminary strategy outline that formally establishes a business relationship with potential customers. Inventors who plan to manufacture and sell an invention without license agreements are strongly recommended to develop prototypes because they will need proof-of-function for distributors and retailers looking to ship or carry the product. Inventors looking for materials and specialty parts should check McMaster-Carr at **www.mcmaster.com**.

Even if for no other reason, a prototype should be built to solicit retailers and obtain high-value feedback to increase the scope of one's market research. The response from a potential licensee is some of the most important information you can receive because it represents the type of feedback your invention is likely to receive from *their* customers. Prototypes should be manufactured in limited runs and probably will cost more per unit than the final version will cost the manufacturer in large quantity. If the cost of a limited run is beyond your budget, consider developing a virtual prototype. A *virtual prototype* is more cost-effective and can demonstrate your conception in three dimensions as a computer-generated, animated model, which can be rotated on screen. To create a virtual prototype, go to Virtual Prototype at **http://virtual-prototype.com**. The fee for creating a virtual prototype is a fraction of the cost of getting a physical prototype manufactured and redesigned.

Finding an industrial designer to help consult on the prototype for your invention is a three-step process. First, you must find an industrial designer who is willing to assist in building your prototype. For a list of designers, check the *Thomas Register*, a multivolume directory of industrial product information at **www.thomasnet.com,** and use keyword "prototype," "engineer," or "industrial designer." The second step once you have located a list of designers is to have them sign a nondisclosure agreement before showing your conception. Explain that you are not looking for a final design but rather a working model. Third, have the designer discuss your prototype options regarding the methods and materials. If possible, present the designer with a two-dimensional sketch of the invention. Get at least three price quotes from three different designers before selecting your designer.

Some notable industrial design firms:

- Creative Edge Products (**www.creativeedgeproducts.com**) provides services such as product development, industrial design, rapid prototyping, investor presentations, patent help, product manufacturing, and marketing.

- Plastic Resource Group (**www.creativeprototyping.com**) works with inventors on a startup budget, entrepreneurs starting a business, and established businesses.

- T2 Design (**www.t2design.com**) offers product and invention evaluation, patent searches, as well as experience in conceptual design, package engineering, and prototype construction.

- Flashpoint Development (**www.flashpointdevelopment.com**) offers target market data, company analysis, product road mapping, computer-aided design, product extensions and platforms, invention and redesign.

- Alpha Prototypes (**www.alphaprototypes.com**) specializes in rapid prototyping, the process of converting 3-D CAD drawings to physical parts to create conceptual models.

- Jacobs Associates, LLC, (**www.jacobsassociates.com**) specializes in product development and manufacturing services, which include prototypes, production tooling, and short-run production.

Creating a prototype

An idea is not an invention. After gathering information on industrial design firms, you will have employ what the USPTO calls "reducing an invention to practice." When reducing an invention to practice, an abstract idea systematically moves toward creating a product that winds up in the hands of consumers. The first step in reducing an invention to practice is making a working model or prototype.

So, how do you built a prototype? If you work in a lab, shop, or garage, you may be an experienced model builder. If you are an engineer or designer, you may have employer resources or manufacturing contacts within your field of invention. On the other hand, some inventions may require only a combination of easily assembled parts that are readily available in stores. Others may be assembled through customized parts. If none of these situations apply, finding a firm to design a model or prototype may be necessary. *See the previous section for a listing of firms you can contact.*

If your invention is a garment, you will need an industrial designer who can develop samples. If your invention is a device, you will likely want to work with a firm that creates some type of mold and pours or injects the mold with a substance that then hardens into your invention. Products made out of hard raw materials, such as metals and plastics, can have prototypes created from polymers, resins, silicone, urethane, plastic, or many other engineered substances that can be poured or injected into molds or built up through layering and curing processes. This reduces the need to tool expensive molds for a single prototype.

Prototype building requires finding the best construction material that allows you to fabricate one or a few models inexpensively and quickly. Although prototype construction is an investment, you need not purchase molds or other products that are substantial and refined enough to mass-produce consumer goods, nor do you need the most cutting-edge or expensive technology. All you need is a rough example of your invention, a real, physical object that performs the functions you have conceived.

Look for firms that specialize in inexpensive prototyping technology for your particular industry. If you have invented a medical device, you want an industrial designer who will create a model from medical-grade materials. If yours is a device intended for use in the ocean, you need a designer who is familiar with materials that stand up to the eroding properties of seawater. Does your invention have flexible parts or have to withstand high heat or strain? Look for firms that talk about their capabilities in these areas, and ask to be put in contact with previous clients. Traditional prototyping methods include:

- **Casting**: The process of pouring liquid into a mold. When the liquid hardens, the mold or cast is taken away, leaving a metal or plastic part.
- **Fabrication**: The process of manipulating a sheet of metal or plastic into parts and assembling the parts
- **Machining**: The method of filing away at a block of material to create a part

Newer methods of creating physical prototypes include more modern methods of fabrication, machining, and rapid prototyping, where a digital model is given physical form by laying down and curing thin layers of resin, powder, or sheets of material until the model is built up.

Your goal is to obtain the type of prototype you need to refine the invention while keeping costs down. A prototype is not intended for mass production, so expensive investment in mass production-quality molds, tools, or other parts is unnecessary. If you intend to manufacture large runs to build your own business without a licensee, identify a manufacturing partner when the project is scaled up.

When you contract an industrial designer, expect to pay a retainer equaling one-third of the estimated cost up front, another third upon completion of construction, and the final third when you receive the prototype. Have the firm provide you written documentation of any changes you discuss along with adjusted costs. Targeting industry-specific designers using the Thomas Register can be essential to the design of your prototype; they can provide knowledgeable design suggestions. If they offer consultation, make sure they cannot claim themselves as a co-inventor with intellectual property rights by reading the fine print in the contract you sign. If the contract specifies that any work performed is strictly work-for-hire, you still own all the claims.

How many prototypes you need to build depends on two things: the cost per unit and the number of prospective licensees targeted by market research. If the invention is not a proof-of-function prototype, you may not want to invest in more than one unit if the invention is designed only for show. If you are creating a **"fit-and-finish" prototype** (in other words, a prototype that represents a final version and requires no redesign), and have enough packaged prototypes made for each prospective licensee, consider sending them to magazine editors who profile new products in your industry.

Additional prototype assistance

University programs and government agencies may be able assist your prototyping needs. The U.S. Small Business Administration's SCORE program offers referrals to retired industrial designers who have experience building prototypes for large corporations. To find a local SBA office, go to **www.sba.gov**.

State or national manufacturers' trade associations also provide lists of members with information about their specialties. Local trade schools and vocational and technical colleges have instructors with industry and classroom experience in manufacturing techniques. Four-year colleges and universities often have technology assistance centers designed to provide cutting-edge advice to manufacturers in the state where the school is located. Many inventors work for universities in their research and development departments, so ask to be put in touch with their inventors, and explain you are looking for advice. The Association of University Technology Managers at **www.autm.net/Home.htm**, for example, is a nonprofit organization composed of university inventors dedicated to creating university-developed patents to be transferred to the private sector. To find inventors who work at universities, call a select university and ask for the research and development department or technology licensing office. If you desire government assistance, the USPTO offers the Inventors Assistance Center — staffed by former supervisory patent examiners with previous experience as direct managers to patent examiners — that provides a list of inventor associations. A list of inventor associations by state can be found in The Blatant Opportunist at **www.tinaja.com/glib/invenorg.pdf**. International inventor associations can be found with Inventnet.com at **www.inventnet.com/international.html**.

Modifications

After demonstrating your prototype's proof of function, industry experts might question the marketability of your invention for any number of reasons. Perhaps the market is not big enough or is too crowded, or maybe your invention has design flaws. In some cases, a potential licensor may think your commercialization strategy will leave them vulnerable to adverse competition. It could be that the product is more difficult or expensive to manufacture than you realize or requires more capital than the firm has budgeted. The industry might be moving in a different direction and pursuing a new or alternative technology that leaves little room for your invention. Perhaps your invention is not distinct enough from the firm's existing product line or is too distinct from it. You might even learn that the current application you have in mind for your invention is not realistic. To pitch your invention successfully, sometimes you must set aside your ego and be willing to modify the original version of your product according to what you discover from speaking with members of the industry. When modifying the original design of your invention, keep in mind that you are modifying it for the customer. The inventor's customer is the manufacturer, not the end user, so listen to what they are saying, and modify according to the feedback they provide.

Packaging

You can further develop marketing efforts by demonstrating a vision of the product's packaging to prospective licensees. When the prototype is complete, locate a professional packaging manufacturer or printer and obtain sample packages from them. A sample does not require artwork; it simply must be packaged the same way products similar to yours are being packaged. To find a packaging firm, ask manufacturers during interviews if they can identify the packaging firms they work with and collect their contact information. When contacting these packaging firms by e-mail or phone, describe your product, and ask for an estimate on the number of units you are interested in packaging. Be sure to have the dimensions of the prototype handy when contacting them.

Building Connections through Industry Research

The purpose of doing market research is to map out a marketplace for your invention so you know where to solicit business. The overall business topography of your invention's potential industry is called the **distribution tier**. The distribution tier is the path a product travels in order to go from being manufactured to being available on a store shelf. The second purpose of market research is to provide the businesses you are soliciting with concrete evidence your invention has a tangible market. This chapter will focus on mapping out the marketplace. *Chapter 5 deals with soliciting manufacturers and distributors.*

Understanding Your Product's Value

What is your invention worth to you? More important, what is it worth to a potential buyer? To strike a deal with a licensee or buyer for your invention, you must be able to gauge your invention's worth. The estimated worth your invention has in the eyes of the manufacturer in terms of its potential to increase profits, its ability to capture market share, and the existing demand for similar products is your invention's **perceived value**. You may see your invention in terms of its design and its features, but according to many marketers, a customer sees a product for the benefits they believe it will provide. Marketers sell customers on the benefits of a product because its benefits are what the consumer perceives as the real value. When considering your market research, examine similar products and begin to research and identify their perceived value. Ask yourself, what makes that perceived value a critical factor in their market success?

Determining Market Position

As you continue to make new connections in the industry, you will come across people who see the value of your product but do not believe it is right for them. To avoid connecting with the wrong people, the independent inventor must be able to determine the marketing position of a company before approaching its key decision makers. This can be achieved by asking members of the sales force if the product you are offering supports the company's market position. The types of products and services offered, reputation, and share of the market they command defines a company's market position. If the company representative being interviewed indicates that your product is not in line with their marketing position, they may have valuable insight into the industry you are trying to sell in, but you are not likely to be doing business with them. As a last resort, ask the interviewee to discuss how willing the company might be to force their market position into the segment that aligns with your product.

Determining market value

Market value is the price at which a seller is willing to sell a product and a buyer is willing to buy it. **Retail market value** (price set by the retailer) is based on the total cost of materials, manufacturing, packaging, distribution, and marketing. When pricing an item, retailers take into account the cost to acquire the item plus minimum profits and the maximum price the market will bear based on perceived value of the product. The purpose of conducting market research is to determine the perceived value of your invention to purchasers on the market. Therefore, what you need to extrapolate from perceived value is the actual value (i.e. market value) of your invention to your customer, the manufacturer. To find the actual market value of your invention, visit stores that carry similar items or shop online. Select a few items to benchmark prices. How were these prices calculated compared to the product you envision being developed in terms of material costs, manufacturing complexity, packaging, and marketing requirements? Will your product be cheaper or more expensive to produce? These questions can be asked during interviews with key people in the distribution tier. The answers are your guide to formulating your invention's market value.

The next step is to compare your invention based on features and benefits. Do the competing items have the same functionality as your invention? Do they provide the same benefits? Is your invention a significant improvement over these items? Compare the prices of the benchmark items you have selected. Are the prices similar? Based on your comparisons of costs, features, and price, use your judgment to estimate a price you think the market can bear.

Consumer Research

When mapping out the marketplace, it is easiest to begin by listing all the end users. If the end users are consumers, make a list of retail stores where consumers buy similar products, then contact the retail or sales managers, and ask to interview them. Retail and sales managers will be able to share important consumer insights. If the store retail or sales manager agrees to an interview, set up a time at his or her convenience. During the interview, come prepared with a list of questions that will help you learn about the industry you are inquiring about. Ask what they like and dislike about the products you have deemed to be similar to yours. Ask about competing entities in the distribution tier. Ask to see consumer sales reports, what consumer trends they see, and whether their customers compare prices or read labels. During this interview process, compile a list of companies within the distribution tier and use that list to begin setting up interviews at the distribution level.

The key to developing interview questionnaires for the supply chain is to begin with specific questions they are likely to answer and then move toward broader questions before closing with open-ended questions. This allows you to build rapport with your subject before asking them for their professional insights and sales numbers. Do not be concerned if you are not sure what questions to ask sales managers and retailers. The more people you interview, the more insight you will gain into the industry, the more questions you will be able to draft into your questionnaire. Some basic questions you can ask include:

- What are the names of the reps that sell the products?
- Which manufacturers seem the most dominant?
- How much does the price of such products affect sales?
- How often do they reorder the products?
- Are there mail-order catalogs within this industry?
- What trade associations serve this industry?
- What consumer trends do they see?

Working backward

The point of starting at the store level is to work backward through the distribution tier, from the store manager to the distributor's sales force until you have reached and interviewed the manufacturers who made the products that are similar or related to your invention. From the store's sales and retail managers, you will get the contact information of the distribution representatives. For the most part, your ever-expanding questionnaire will apply the same questions to retailers and distributors alike, but keep in mind that retailers will have more localized industry information whereas distributors will have more information regarding an entire region. Questions for distributors also should be geared toward learning about the industry's manufacturers. They should provide insight into:

- The pros and cons of working with certain manufacturers
- The market shares these manufacturers command
- Their presence in foreign markets
- Which manufacturers have the most licensing agreements with independent inventors?

After your interview with distribution representatives, you will want to obtain the contact information of the manufacturer representatives they deal with. After you have interviewed the manufacturing representatives, your goal is to obtain the contact information of their superiors. These will be the key decision makers inside the manufacturing firm, such as marketing managers. Key decision makers inside manufacturing firms are the people you will be soliciting a license to. The path to a licensing deal is market research; therefore, you should be using feedback to improve your invention to meet the needs of the market better.

Contacting the Sales Force

No matter how much valuable information about the industry you receive from retail managers at the store level, the purpose of your interview hinges on getting the contact information of the sales force that sold similar products to the retailer. The sales force that represents the distributor will be able to connect you with the sales force that represents the manufacturer. The sales force that represents the manufacturer will be able to connect you with the key decision makers who negotiate licensing deals with independent inventors.

The first step in contacting the sales force is to ask if the person you are talking to is a company salesperson or a sales representative. The distinction between the two is that a company salesperson sells only the products of the company he or she is working for, whereas a representative works for multiple noncompeting manufacturers or distributors. Although a salesperson may have greater insight into the company you are inquiring about, a representative may have less bias and more knowledge of the overall industry. Which company each sales force serves may influence their perspectives on the marketplace differently. Therefore, what type of sales force a company uses may tell you a lot about the company's business model and how they conduct business. For example, larger manufacturers prefer to have tighter governing control over sales decisions and therefore, tend to employ their own sales force. Establishing familiarity with the type of sales force each company uses and knowing the way they conduct business can help you decide which company could best serve your needs.

When contacting someone in the distribution chain, always mention the name of the previous person in the distribution chain who provided his or her name as a reference. Providing a reference lets the new contact know that one of his or her customers in the chain was interested enough in your product to offer the name and contact information of the person they directly do business with. As a result, he or she will be more likely to devote more time to an interview and will be more inclined to hear about your product. The timetable for conducting interviews for market research can run anywhere from a few weeks to a few months. If time constrictions or reticence prevents you from doing the work it takes to compile market research, professional companies that specialize in this service are available for fees between $1,000 and $10,000, depending on the scope of the work required. *For more information, Chapter 10 provides related information, including how to avoid marketing scams.*

Pull-Through Sales

Having a prototype developed to show your contacts is a critical way of gauging initial interest during your market research interviews, and at some point after the interview process, you may have to test-market the prototype to demonstrate pull-through sales as proof your product has a market. **Test marketing** is the method by which a prototype's sales potential is measured and provides some insight into how you can most effectively achieve the kind of sales your customer is looking for. **Pull-through sales** are the amount of sales an individual company estimates a product must have in order for it to be considered marketable. The more marketable the prototype, the more licensable the prototype will be considered. The purpose of conducting interviews at this stage is to learn about the industry, as well as to identify which potential business partners you will consider soliciting licenses to later on. Therefore, a good way of determining this is to ask each contact what type of pull-through sales he or she would require from your product in order to consider licensing it. If you achieve or even supersede his or her expectations, your product will be considered. If you test-market your prototype and it underperforms, he or she most likely will not consider your product.

Test-Marketing

To effectively test-market your prototype, you will need a budget that covers appropriate packaging costs and the amount of production models it will take to demonstrate pull-through sales. To test-market a product, marketing services will ask for fees that range from $10,000 to $50,000. If you handle the test marketing yourself, you will need a budget that covers the cost of advertising, press releases, events, incentives offered to salespeople who can get it to store shelves on a limited run, focus groups, and so forth. However, if you do not have the budget, there are two inexpensive ways to test-market and still effectively determine pull-through sales. The first is to create a simple website for your product and send targeted traffic to that site through Google AdWords for under $500. Using your product's traffic statistics in Google Analytics, you would be able to adjust the time and frequency for when you would like the advertisement to show to achieve a better flow rate. In fact, before you even design a prototype, consider creating a simple website as a survey tool to ask people if they would consider buying your product and how much they would be willing to pay. The second inexpensive test marketing strategy would be to try selling it on eBay through an auction.

Whether you choose to test-market online or through retail stores, remember that large retail stores are more inclined to rely on regional sales figures from stores over online sales in the test market as the basis for the decision to license a product. To create a regional test market, approach independent retailers whose managers have the ability to make immediate purchasing decisions, and over a period of weeks, monitor your product's sales performance. Consider having alternative designs and packaging made to determine which option the buying public is most responsive to. Adjust the pricing to see what amount begins to increase and decrease sales. Offer different prices in different stores and compare the sales figures. In some cases, a potential licensee will not require proof of pull-through sales, so when you reach the manufacturers during the market research phase, ask them if they require sales figures from a test market or if they are willing to facilitate a test run. Some of the more experienced key decision makers will be able to determine sales potential without a test run. However, if you are an independent inventor, you probably will want to know your product's sales potential before anyone else can make a determination, so if you test-market your product, compare your sales to the overall market for similar products. Whatever percentage you command regionally over a period of weeks or months will give you an idea of how many units you could potentially sell over a year.

Production Costs

The easiest way to estimate your invention's production costs is to supply a manufacturer with a complete list of components and materials needed to build your prototype and ask for a quote to produce 10,000 units. These costs will have two components: fixed and variable. Fixed costs include tools, molds, or other equipment costs that will be incurred regardless of how many units you order. Variable costs are reduced as orders get larger because of discounts on materials and economies of scale. Your cost-per-unit is lower as you make more products not only because of the lower price per unit from variable costs but also because of the reduced share of the fixed cost represented in the cost of producing each unit. Prospective licensees will know or be able to find out easily through their suppliers what it will cost to produce your invention. Because manufacturers work with established suppliers and do much of the production work themselves, they will wind up with better quotes and lower production costs than you would by simply cold calling suppliers for quotes.

New Versus Existing Products

Companies also introduce new products to increase their market shares, respond to competitors' innovations, take advantage of new technologies, or reduce costs. According to professor Ken Homa at the Georgetown University McDonough School of Business, about a quarter of the time or more, new products are often improvements on existing products. When a company is making an improvement to its existing product line, it is unlikely to be licensed from an independent inventor.

Another quarter of new product introductions are known as line extensions. **Line extensions** are new products that complement existing products. For example, when Mattel® introduced the Barbie® doll, it followed with different types of clothing accessories, followed by Barbie's camper and her dream house. An independent inventor could have invented and patented something Mattel could have used in the Barbie line extension if Mattel felt it added to their market positioning in the toy doll industry. Line extensions currently represent the largest segment of opportunities for independent inventor submissions.

About 20 percent of all new product introductions exist somewhere else, which means a company is licensing rights from an inventor who already has multiple license agreements in different industries or geographic areas, or the company has decided to introduce an identical product based on non-patentable ideas. Only 10 percent of all new product introductions are truly original products. These are not considered spin-offs of an existing product, new accessories for an existing product line, or retooled old products with new price tags.

New product risks and motives

According to professor Homa, only 30 percent of industrial products and 15 percent of consumer products manufactured wind up succeeding in the marketplace. Statistics indicate that although new product development is necessary for companies, it remains a risky enterprise. Of the small percentage of new products that become commercialized, even fewer turn a profit, hence the reason why many risk-averse, profit-focused corporations continue to build on proven successes rather than seek original innovations.

So, given the inherent risks, why would a manufacturer want to introduce new products? The most common reason is this: Manufacturers looking for new and unique innovations are the ones looking to aggressively capture large portions of largely untapped market share, and new product innovations are the best way of accomplishing that. When pitching your invention to a manufacturer, you must consider several factors weighing in its decision to license your idea.

The first factor the manufacturer will consider is your invention's perceived value. The higher the perceived value, the more likely a manufacturer will offer a licensing agreement that turns your invention into a product that can be distributed. Once you have reached the key decision makers, you will want to increase the perceived value of your invention by presenting your own market research. If the manufacturer is sufficiently impressed, it will then conduct a critical evaluation of your invention using its own market research to determine the risks versus the benefits of acquiring the rights of your invention.

Other factors influencing perceived value include whether the new product fits well with the existing line and the degree to which the manufacturer already has a market presence in the industry your invention will be sold to. If it is already good at selling this type of product, it has already established itself in the type of distribution tier that can get your product to the end user and at a cost that will turn profits. If it has a significant market presence established, it knows who the invention's potential consumers are in its sales base and which customers will consider new product introductions. Your invention can help the manufacturer shore up or increase its market position.

Another critical factor manufacturers consider when evaluating your invention is production costs. If you have successfully pitched your invention to a key decision maker, the manufacturing department will determine what the cost will be to convert the invention into a product it can sell to a distributor. By determining the production cost, it is able to determine the selling price of the invention. The difference between cost of making your product and its selling price is the product's **gross profit margin**. Once the gross profit margin per unit is determined, the manufacturing department then uses its market research to estimate sales volume and life cycle to extrapolate the overall value your invention will have to the company if it agrees to a licensing deal.

Manufacturing firms have to perform a complex calculus before deciding in a new product's favor. Not only do profits have to exist, but they also have to be high enough to overcome the inertia created by the startup cost plus the risk of new product failure. Most products do not end up profiting enough to compensate for the startup costs and risks associated with a new product. If you can position your product to appeal to your customer's brand and financial positioning strategies, you have created powerful motivation and may hook your customer on your invention.

The Size of Your Market

To know how many units to produce during the first manufacturing run, you have to be able to estimate the size of your market. There are economies of scale in manufacturing — runs of higher volume equal lower cost per unit manufactured. Census data is one resource for identifying how many potential customers exist for your product. The Census Bureau publishes statistics at **www.factfinder.census.gov**. To determine which specific groups of people would use your product, ask yourself who would most likely benefit from what your product offers. Interviewing retailers, salespeople, and key decision makers up and down the distribution chain of a product similar to yours should further narrow down your customer demographics. If you can define your customer demographics — (for example: women over 65, parents with more than two children under the age of 18, men 16 to 25 years old, households with income over $60,000 per year, and so forth) — data from the U.S. Census Bureau is available on how many Americans fit those criteria.

By capturing demographics, it may then be possible to discover what percentage of that demographic will actually purchase your product. Your market may be segmented. Public companies publish annual reports of market segments that contain data about the size of market segments. Annual or quarterly reports can be obtained with a phone call to company headquarters or through 10K filings published on online databases that offer company profiles and industry information. One of the most popular company profile databases is Hoovers™ at **www.hoovers.com**. Public libraries also may subscribe to business magazine and journal services such as ProQuest or InfoTrak. You can also search for articles pertaining to industry market segment and analyses from shareholder meetings transcripts or podcasts on the Internet.

The greater the number of competing products within a market segment, the smaller your market share is likely to be. Websites such as InventionCity. com estimate that proposed market shares of between 5 and 20 percent will seem realistic to prospective licensees most of the time. You will have to make a strong case to prospective licensees for why your product would capture a market share above 20 percent. They also advise falling back on 10 percent as a reasonable answer when you have no clue about market share.

CASE STUDY: COMMERCIAL POTENTIAL

Jack Lander, president
Inventor-mentor.com, LLC.
949A Heritage Village, Southbury, CT
06488
jack@inventor-mentor.com
www.inventor-mentor.com
(203) 264-1130

Professionally, I claim several job titles as a manufacturing engineer, product designer, inventor, author, and columnist for *Inventors' Digest* magazine. As past president of the United Inventors Association — a non-profit dedicated to helping inventors — I know something about evaluating inventions. The evaluation service I run for clients offers an evaluation of their invention's commercial potential. About 90 percent of the inventions I see have little commercial potential from their original conception, so at the beginning, I work to highlight all the weaknesses for the inventor to try to resolve.

When I see the commercial potential in an invention, I work to help my clients refine and prototype. Getting the conception right for a prototype is critical; otherwise, the inventor winds up wasting money on expensive molds or other production-related investments. When I do an evaluation, I work to verify that a viable market actually exists for the invention. Afterward, we examine what type of patent is needed and where the industry's manufacturing customers can be found.

I strongly advise inventors to pursue catalog companies because they are the most inventor friendly. Chain stores predict their return on investment for every square inch of shelf space. They do not want to gamble on products that lack a sales record because their calculation of ROI is based on market projections. Until you can show you have a reliable sales record and inclining sales, executives at chain stores are reluctant to take on your product. Chain stores want product lines; they typically do not like to do business with someone selling a single product.

Catalog companies, on the other hand, are willing to deal with a single product. They want innovative products not found anywhere because that is their niche. They do not request inspections of your facility, so if you operate out of your basement or garage to keep the overhead low, it will not be a problem.

If clients want a prototype that looks like the finished product, they will probably require industrial design assistance, so I refer them to established industrial designers who can provide design input. Building a prototype depends on what you are looking for. In some cases, prototypes should be designed for functionality rather than appearance. In other cases, their appearance is more important.

I also advocate using a sell sheet when advertising the product to prospective licensees and distributors. A sell sheet should describe your conception; emphasize key benefits; and include photos, illustrations, and testimonials. Some inventors contact me when they already have a patent, others when they only have the idea. In most cases, it is not necessary to spend any money on obtaining a patent or a prototype before doing an evaluation. One time, I had a client who invented a product to protect bras in the washing machine and decided it was so revolutionary that the client obtained a patent before contacting me. As it turned out, the inventor had no evaluation to base his finances on and was unable to afford the tooling to produce the invention at the time the patent was issued. A competing company saw the patent and legally found a way to design a similar product around the patent. The competitor's product was the first one to market. My client had a higher quality product but, unfortunately, got locked out of the catalog markets because of his mistake. The most important thing is to have a marketability or commercial potential assessment done early in the invention process.

Identifying Your Customers

Profiting from an invention means targeting the right customer, and a complex set of considerations are involved in doing this. If you lack the capacity to create a product and deliver it to the end user — the consumer — you will need the help of someone with the proper resources. If so, your customer is not the end user, but rather, the manufacturer.

Targeting the Manufacturer

The end user who purchases your invention in the store will likely never know your name; therefore, your target customers are the key decision makers inside the manufacturing firms who specialize in creating products that relate to the industry your product will be sold in. Key decision makers tend to be managers and vice presidents who run the firm's marketing department. They are the ones with the power to grant an evaluation of your invention, pursue licensing deals with inventors, and approve business partnerships. Without this partner, your invention may never be a product on a store shelf.

To find the manufacturer of any product, check a product's packaging to see if the name of the manufacturer is listed. If not, ask the store manager which distributor or wholesaler delivered the product to the store. Finding the right decision makers means identifying the right companies, and this may require having to retrace the various points where the product changed hands. To retrace the distribution tier, use market research to decide which industry your invention should be sold in, develop a list of retailers who sell products related to that industry, approach the retailers on your list, ask them who delivered the product, and work backward to find the previous source in the distribution tier until you have located the manufacturer. The retailer is anyone who sells the product directly to the end user in his or her store. They include:

- Discount stores
- Specialty stores
- Department stores
- Convenience stores
- Automatic vendors
- Off-price retailers
- Mass merchandisers
- Warehouse clubs
- Corporate chain stores
- Retailer cooperatives
- Consumer cooperatives
- Franchises
- Merchandising conglomerates

Once you have identified the size of your invention's potential consumer base, it is time to figure out how you might get your product into distribution in its industry so it can reach the retail venues where those consumers buy. For your product to reach consumers, you will begin talking to your actual customer — the manufacturer.

To generate a list of manufacturers, consult the *Thomas Register*, which is available in libraries or online at **www.thomasnet.com**. After identifying your potential list of customers (manufacturers), you will need to learn everything you can about each manufacturer on the list. Doing this will not only provide a clearer sense as to whether the manufacturer might be a viable suitor for your invention but also provide a broader understanding of the marketplace and the viability of your invention within the industry it serves. Learning about these manufacturers should indicate what types of licensing deals they seek with inventors, the scope of their terms in the licensing deal, and their monetary success rate with inventions they acquire through licensing agreements. Another way to locate a target list of manufacturers is through an online database search. According to **www.manta.com**, a free database on company profiles and industry information, there are more than 31,000 manufacturing companies in the United States. For a fee, Manufacturers' News Inc. offers an expansive database of manufacturers broken down by region and industry, the latest startup manufacturers, and a comprehensive executive database that provides the contact information of key decision makers. Their website can be found at **www.manufacturersnews.com/database.asp**.

As you scour your list of manufacturers, it will be important to consider that all companies strive to behave in a way that is consistent with their business models. A company's business model is the description of its reasons for creating specific organizational structures and certain operational methods in order to create value. Company executives make decisions that are aligned with this model because they have had previous success in adhering to it. When you consider approaching a manufacturer, ask yourself if your invention matches its business model.

As previously discussed, you should be able to determine if your invention matches a company's prospective business model by speaking with its sales force. A good business model is one that constantly adds value to the company's gross profit margin. To determine the strength of a business model (i.e. the rate of its gross profit margin), gather industry reports from a related trade association (*discussed later in this chapter*) and make a list of at least five competitors in the same industry. Then, determine the sales generated by each competitor by reading their annual 10K reports. Using their 10K, find the cost of sales total (or add them up by looking at raw material costs, inventory, shipping, direct labor, and other costs associated with sales). Keep in mind that cost of sales does not include cost of operations, such as rent, administrative labor, or utilities. Once you have the total cost of sales, subtract that number from total revenues from gross profit (also listed in the 10K). For example, if a manufacturer's total revenues are $20 million, and its cost of sales are $10 million, and then its gross profit is $10 million. Then, calculate the gross profit margin by dividing gross profit by total revenues. For example, $10 million divided by $20 million equals 50 percent. Whichever company has the highest gross profit margin has the best business model because its model is creating the highest value or rate of return.

Working with Licensees

Once your invention has been tested and evaluated by a manufacturing firm, a key decision maker in the company will review the analysis and make the decision about drafting a licensing agreement. Key decision makers in most industries will be one of the following: company owners, presidents, marketing VPs, general managers, or buyers. By interviewing these key decision makers for your marketing research, you not only are compiling information about the industry but also determining which companies match your business goals. Before contacting them, decide whether they are the best prospect. Your best prospective licensee is one that:

- Has cash from high profit margins
- Lacks debt
- Is looking for a niche
- Is inventor friendly
- Sees your invention as complimenting or enhancing their market

Companies have an incentive to look for outside licensing deals for several reasons. First, most licensing agreements allow companies the leverage of retaining licensing rights or handing them back to the inventor if they choose not to produce the invention or if the invention does not post a profit. Second, most companies do not pay independent inventors cash up front, which allows them to free up capital for marketing purposes. However, if the company has the rights assigned rather than licensed, it means they permanently own the rights and will have to pay the inventor money up front for the sale of all rights.

So how do you determine which ones are the best potential licensees? Some inventors start with small firms they believe are more accessible. Others start with large firms that may have more resources and better success at pushing new products. Suppose your market research found three dominant industry-related firms, each with 20 percent control of the market for products such as yours. One of these firms, however, may be owned by a multinational corporation that sells many brands in multiple industries and may have more capital to allocate than the other two with equal market share. The drawback of licensing to big firms is that they may have more attorneys who can help them design around your intellectual property protections or create provisions that tilt the licensing contract heavily in their favor. A large firm may have in-house research and development staff that may not be outside-inventor friendly. It can be slow to make decisions due to many layers of bureaucracy. Although you may hold an advantage with respect to these issues when dealing with smaller firms, they may have less marketing capital, command a smaller market share, and are more at risk of declaring bankruptcy after you license to them. Because an assignment grants the permanent transfer of rights from the inventor to the manufacturer, and a license agreement grants a temporary transfer of rights lasting the duration of the agreement, your decision to license or assign your rights may depend on whether you want to make more money up front in an assignment deal, or license and retain your option of doing business with someone else at some point in the future. Be aware, however, that if you license to a manufacturer, the manufacturer has the rights to sublicense or assign its temporary rights to another manufacturer or distributor to make or distribute your product. If your manufacturer sublicenses the rights to someone else, the original manufacturer is still the party responsible for overseeing production and paying you. If the original manufacturer assigns the rights you temporarily licensed to them, the original manufacturer has absolved itself from all rights and responsibilities and transferred them to the new company that is now responsible for paying you royalties and overseeing production. Most inventors do not like being blindsided by a licensee who assigns its temporary rights, and therefore, these inventors stipulate in their license agreements that their temporary rights are nonassignable.

Waiving rights to the invention

Companies, particularly large companies with readily available legal resources, will require an inventor to sign a waiver that relinquishes a number of important rights to the invention. Waivers are designed to protect companies from suits that may arise and claim that the company violated some implied confidentiality agreement or violated an implied agreement to pay for using the invention or some specific part of the invention. In the past, many companies have lost these types of lawsuits, not so much because of the terms of the agreement, but because of the time, expense, resources, and uncertainty of litigation. Even in situations where the company's own inventors were responsible for developing an invention, independent of any outside inventor, companies have lost these types of lawsuits or settled for a compromise rather than deal with uncertainty and high cost of litigation.

A waiver requires that the inventor give up rights, but it may not force the inventor to relinquish any rights under patent laws. A typical waiver will require the inventor to agree that the company is not obligated to pay for using an idea, to keep the idea of the invention confidential, to return any documents submitted, and to have no obligation to the inventor except as established by patent laws. Many companies also add other minor provisions. In essence, a waiver relinquishes all rights against the company except to sue for patent infringement when a patent is acquired.

The usual procedure for most large companies is for the inventor to send correspondence to the company that introduces the invention or idea. The company usually forwards the correspondence to an appropriate individual in its patent or legal department.

The patent or legal department will respond to the inventor by delivering a form letter, which details the company policy and requests that the inventor signs a waiver before the company agrees to review the invention. Upon receiving a signature to the waiver, the invention submission is forwarded to the company's engineering department. There are several issues with this process as follows:

- The patent, if not already issued, may not be issued.

- The company may use a variation of the invention that may not be covered by the issued patent.

- The company is not bound to keep the invention confidential.

It is because of these issues that an inventor should avoid large companies and waivers and concentrate on smaller companies that do not require a waiver. In fact, it is more advantageous for inventors to seek companies that are willing to sign an agreement drafted by the inventor. An agreement drafted by the inventor is known as a Proprietary Submission Agreement. If an inventor is required to sign a waiver, the pending or acquired patent is still protected under patent laws regarding first rights. The inventor must be sure to choose a reliable and fair company and insist that a decision be made within a reasonable time, usually six months, or all documents must be returned to the inventor. Without an agreement with regard to making a decision, a company may takes years to make a determination, which may interfere with other efforts to market the invention.

Proprietary Submission Statement

A Proprietary Submission Statement is an agreement that stipulates a company will agree to review an invention, keep the invention and all attached documents in confidence, to return all documents submitted, and to pay a reasonable sum and royalty if the invention is adopted. The requirement to pay reasonable fees and royalties should include language that specifies the amounts of such funds be settled in the future through negotiation or arbitration. If a company refuses to sign, the Proprietary Submission Statement may be modified to eliminate requiring payment of fees and royalty or eliminate the requirement to keep all information confidential and allow the company to disclose any supplied. If these modifications still do not result in signed contract with a company, the inventor may have to rely solely upon having the company review the invention without a contract and settle for reasonably strong rights and protection as they apply under patent laws.

Until you examine your prospects, it will be difficult to know which of them is best. Do your prospects have cash? The best way to assess their financials is by examining their annual 10Ks or quarterly 10Qs, which can be found online at Hoovers (**www.hoovers.com**), ProQuest (**www.proquest.com**) or InfoTrac (**www.infotrac.net**). Do not assume that a division of a large company has tons of cash on hand. They may be dealing with litigation issues or seeking debt financing. A company's 10Ks also track private placement of public equity (PIPE) deals, which means the company will be receiving a huge infusion of cash for future business operations. Conversely, investor activity on the stock market may provide insight into a negative situation, as many stock speculators short-sell as a way of making money on the expectation that a company is going to lose money. Above all, be sure to scrutinize product lines and ask yourself which company's market position is most complemented by your invention. Does your product help address competition, extend a product line, or offer other strategic advantages?

Obtaining a meeting

When you are ready to make some phone calls, make sure you have developed a pitch. Pitching your idea in person is more effective than pitching over the phone or by mail. If you cannot meet in person, request a teleconference with a webcam and an Internet long-distance call.

Having a prepared script will help to communicate information quickly, concisely, and professionally. The more competent you sound, the more confident your prospective customer will feel in dealing with you. Although you need impartial feedback, you also must protect your idea from being known by competitors who would try to produce your idea first. You do not have to reveal the exact details of your invention at this point. Simply describe its main functions and benefits and how it differs from others on the market. Be sure to include some facts and statistics from your market research, including information about the company that led you to believe your product would be a good enterprise for them. Another effective way to prepare relevant information is to consult the DIMWIT™'s Guide for Inventors at **www.dimwit.com**. Dimwit offers advice for invention development and confidentiality agreements, and provides additional online inventor resources.

Decide from the beginning whether it is better to approach one prospect exclusively or to pitch several at once. If the industry is extremely secretive and competitive, it may be better to approach one business at a time and offer each an exclusive evaluation of the prototype. By approaching firms one at a time, you can refine your pitch. If your invention possesses no trade secrets, approaching all companies simultaneously can speed the process of getting an agreement. If your invention is pending patent approval and contains trade secrets, ask the firm to sign a nondisclosure agreement.

Nondisclosure agreements

A nondisclosure agreement is a confidentiality agreement between a minimum of two parties who agree to restrict the access and use of shared information, knowledge, or material from additional parties. *Nondisclosure agreements are covered further in Appendix C.* The benefit of drafting your own agreement is that it recognizes your rights to your invention and strengthens your claim on a patent after it has been registered. A nondisclosure agreement may be necessary if just one key element of your invention has not yet been patented. Some companies — particularly large ones — will not agree to sign a nondisclosure agreement because it puts them at risk of accusations they cannot control, such as the possibility of being held accountable if another inventor with a similar trade secret releases his or her information into the public domain.

If you choose to deal with a company who refuses to sign a nondisclosure agreement, reveal only the basic functionality of your idea. When weighing the pros and cons of large manufacturing firms versus smaller ones, also consider the fact that smaller firms are much more likely to sign your nondisclosure agreement. Having a patent attorney draft your nondisclosure agreement is acceptable, though it is not recommended to have an attorney submit the agreement if you are looking to sign a licensee as quickly as possible. Submitting an agreement through a patent attorney only causes delays because the manufacturer will have its attorney examine the agreement, which could lead to a lot of haggling over the terms. Remember, timing is of the essence because patent law states that once you have publicly disclosed your design, you have only 12 months to apply for a utility patent, otherwise all rights to such claims are forfeit and available for anyone to pursue. A confidentiality agreement, however, binds the signatories to secrecy and prevents the clock from starting on your patent claims because the designs are not disclosed publicly.

In some instances, the company will have a nondisclosure agreement drafted for you to sign. Larger firms are more likely to have nondisclosure agreements because they carry larger liability risks and have a legal department to draft documents. These agreements will be tailored to lawfully protect them from being sued in the event that your trade secrets have been disclosed or developed by someone else. Some companies will insist that you sign the agreement; others will be willing to waive it if you explain you are only going to reveal information that has already been patented and exists in public domain at the U.S. Patent & Trademark Office. The basic structure of a nondisclosure agreement includes:

- A Statement of Purpose
- Date
- Liability and Obligation Clause
- Agreement Clause
- Outline of Terms
- Signature Line

Selecting the Distribution Chain

In most cases, when you find a manufacturer willing to license your invention, it will use its existing partners in the supply chain to deliver your product to the end user. If not, you will have to select a distributor willing to move your product out of the manufacturing facility. To identify this distribution chain, begin by thinking about materials, such as what you used to make your prototype. What are the components of your invention made of? Where are these materials available? What types of manufacturers work with these materials? Consider how manufactured goods reach the consumer and whether they are channeled through distributors, wholesalers, big box stores, direct marketers on TV, or by mail. In the process of identifying the distribution chain, it is important to find out:

- Whom they distribute for

- What products they distribute
- Their competitors
- The markup at each stage of distribution
- Where they fall into the overall supply chain
- Their regional and national distribution presence
- Their market share
- Their sales volume
- Industry trends
- Market climate

The best way to assess distributors in a related industry's supply chain is to compare their competencies in the supply chain. According to Tompkins & Associates (**www.tompkinsinc.com**), a leading provider of business strategy and supply chain solutions, several levels of assessment criteria are used to measure the excellence of supply-chain management. They include:

- Enabling technologies (the quality of technologies used to track products, inventory, and payment schedules)
- Supply chain synthesis (the quality and speed of communication among businesses in the supply chain)
- Warehousing (the quality of the storage for products moving up and down the chain)
- Logistics (the quality of the supply chain's plan to move products)
- Manufacturing (the quality of the process and materials used to make the product)
- Organizational excellence (the quality of leadership within each business)
- Maintenance (the quality of keeping a product protected while in transit)

In order to measure the excellence of a supply chain, list each company within the potential licensee's supply chain and contact their business process managers. Ask them to describe how their enabling technologies handle information flows, how they respond back and forth with other businesses in the supply chain, where their warehouses are located, how they store and move products, how fast it takes them to receive and supply, and how their organizational tree has been designed to optimize and run their business. Some companies are in the business of helping businesses evaluate the core competencies of distributors so they can make an informed choice as to whom they want distributing their product. Supply Chain Edge (**www.supplychainedge.com/about-us/index.shtml**) is a company that has created an Alignment and Assessment Process to help businesses determine which supply chain investments will give them the most substantial returns. Distributors are measured by their ability to distribute effectively. Factors that determine this include their ability to track, analyze, model, decide, and act with respect to goods moving through their supply-chain channels.

Nowadays, distributors use high-efficiency business solutions that track the movement of goods in real time. These solutions are then integrated into a data warehouse for analysis, reporting, and data mining. By seeing the real-time movement, distributors are able to model and score these movements to address problems and make better supply-chain decisions.

When identifying distributors for your invention, it will necessary to learn as much as possible about their operational capabilities. Large distribution chains tend to be less friendly to independent inventors, so it will also be necessary to gage their level of receptivity. Small distributors may be more open to independent inventors, but they tend to have less reach. Wal-Mart Stores Inc., for example, requires each of its suppliers to have an office in Bentonville, Arkansas, so if your distributor is not willing to meet this requirement, your product cannot be sold in Walmart®.

Clearly, organizations that have precise packaging, delivery, and pricing requirements are expecting to deal with large, well-organized distributors. That does not mean you should avoid researching products sold in this way; it does mean, however, that you should consider how your product is likely to be distributed and research those outlet channels that are similar to how you plan on distributing your product as well.

Trade Shows and Trade Associations

Trade shows are events where industry professionals and potential customers gather. Although industry professionals gather at trade shows to sell their own products, they are not there to meet with inventors. Manufacturers and distributors spend thousands of dollars to rent booths, set up displays, and pay representatives to connect with the people in their supply chain. They want to profit from these expenditures by acquiring new business, and they only have a few days to make the event worth their organization's time.

Trade shows are where independent inventors have an opportunity to meet prospective customers, review their upcoming product lines, see how they interact with potential customers, and perhaps make a few contacts. If you can manage to do so, there will be plenty of opportunities to chat with representatives, make contacts, ask questions, gather ideas, and deliver your pitch.

Trade shows relevant to inventors include:

- INPEX — Invention & New Product Exposition

  www.inpex.com. America's largest trade show; provides an exhibition forum for inventors looking to solicit business with manufacturers

- IENA — Ideas, Inventions, New Products Exhibition

  www.iena.de. International trade show in Nuremberg, Germany; represents about 37 countries and 800 inventors per show

- The Australian Innovation Festival

  www.ausinnovation.org. Australian-based initiative to foster innovation and economic growth; hosts festivals and offers seminars, expos, and networking workshops

- Gdańsk International Fair

  www.mtgsa.pl. International fair in northern Poland running exhibitions since 1989 for multiple branches of economy

- Taipei International Invention Show & Technomart

  www.inventaipei.com.tw. Exhibition related to the business services sector, including high technology, fiber optics, machinery, construction, safety equipment, environmental protection, pharmaceuticals and healthcare, transport and communications, household electronic appliances, hardware, agriculture, food, textiles, etc.

- The Ultimate Trade Show Directory

  www.tsnn.com. Online resource for the trade show, exhibition and event industry since 1996; operates event database containing data on more than 19,500 trade shows, exhibitions, public events, and conferences

Trade associations typically host trade shows. You must look for the type of trade associations your invention relates to. Trade associations do more than host trade shows. They also provide membership directories and print publications that cover industry specific news. Trade associations typically have a paid executive director and some administrative support staff. You can call them with questions about the industry. The primary responsibilities of trade associations include providing services to members and facilitating communication within their industry. To find an industry specific trade association, start with the three major ones. They are the Federation of International Trade Associations (**www.fita.org**), the

International Trade Administration (**www.trade.gov**), and the Consumer Action Web (**www.consumeraction.gov/trade.shtml**).

Mail Order Catalogs and Online Retailers

After interviewing sales people and working back through the distribution chain to the manufacturer, you should have a good idea of how your invention's industry functions. You should know who the largest manufacturing and distribution firms are and how they operate. With luck, you will have names of individuals responsible for making purchasing decisions within these manufacturing firms. These are your prospective customers — the folks who may license or purchase your invention or partner with you to turn your invention into a salable product. But have you reached out to everyone?

Not all products are sold in retail stores. Mail-order catalogs are considered a secondary market to retail stores, but their value to the inventor can be enormous. According to the National Mail Order Association, mail-order catalogs are a $400 billion industry, and there are more than 11,000 consumer catalogs in the USA and more than 6,000 business-to-business catalogs. The supply chain for this burgeoning business is relatively simple. End users who want to buy products listed in catalogs make purchases from the mail-order catalog company, who in turn purchases the product from the manufacturer who is licensed to make and/or distribute your product. In some cases, if the sales order is lower than the manufacturer's minimum order requirements, the catalog company will purchase directly from the manufacturer's distributor, which is likely to have units of your product stored as inventory in their warehouse. Catalog companies tend to be excellent secondary markets because they are inventor friendly, as new product introductions are a standard feature of some catalogs.

Many familiar consumer products, including the pop-up toaster and steam iron, were first introduced to consumers through mail-order catalogs. The

Hammacher Schlemmer catalog (**www.hammacher.com**) has a contest for certain types of patented inventions that, for example, may address specific lifestyle problems. The contest is free, must meet stated requirements, and is open to the national public. Online entry submission forms can be downloaded by going to their website. Brookstone and Sharper Image also run catalogs that specialize in introducing innovative products to their customers. Large catalog operations such as Miles Kimball, Fingerhut, Damark, Hanover Direct Inc., Lillian Vernon, and The Johnson Smith Company mail millions of catalogs a month to homes all over the country. To reach these mail-order catalogs, search Catalog City online at **www.catalogcity.com,** and peruse the store category section to find a related industry.

Specialty online retailers also have exploded in the past decade. Nowadays, one can find retailers who specialize in everything from animal-themed products to aids for the elderly and disabled. Catalogs.com at **www.catalogs.com** is also an online aggregator of catalog companies. Their site includes traditional mail order catalogs and online catalogs. You may not find catalog buyer information in a mail-order catalog or on the website, and there is not a typical distribution chain to work through because catalogs deal directly with inventors and manufacturers. Instead, check their website for submission instructions or consult a business directory such as the Thomas Register for the contact of a company executive and ask to speak with that person. Another option would be to consult a specialized directory for the mail-order catalog industry. Two of the largest directories in publication are the Grey House Publishing's Directory of Mail Order Catalogs (**www.greyhouse.com/pdf_dc/dmoc_dc.pdf**) and the National Mail Order Association's National Directory of Catalogs (**http://nmoa.org/**). For a $99 fee, the National Mail Order Association offers a service to inventors providing access to a database that catalog companies use and features your invention in a newsletter distributed to their 8,000-member network, which includes manufacturers.

Research and Development Departments

Most products have a certain life cycle and reach a point where they need to be improved in order to compete in the marketplace, or they will be replaced entirely by something different. Because of this impending life cycle, many companies are forced to innovate continually if they want to stay in business and compete. As a result, they are continually introducing new products and looking for new product ideas from their research and development departments or externally to inventors who own the patent on a product they can use. Even the largest, most successful companies in the world have not rested their laurels on maintaining a single product over time. Coca-Cola Co., for example, leads the entire soft drink industry, yet they are constantly introducing new formulas in the market and spending millions of dollars on research. The best way to arrange a meeting with your licensee's research and development department is to contact the department manager or CEO and ask them to act as a liaison. This person knows the schedule of that department and will be able to arrange a meeting with the staff accordingly.

Ruling yourself out

No company will spend money to acquire an invention that does not expand its business or will not turn a profit. Remember, your customer is not the consumer, but rather the company that will produce or distribute your invention. Companies will introduce a product that replaces an existing product only if it is likely to sell more units, save them money, or increase their competitiveness. The benefits an invention must offer its customers should include:

- A substantial improvement in the way the product works
- Substantial improvements in consumer ease of use
- Manufacturing savings

- A classic, sleek, cute, or stylish design that practically sells itself to the customer

Inventor traps to avoid

Your prospective customers may have a limited amount of capital and a variety of new product ideas to consider. They will be choosing among competing inventions and product ideas in deciding where to put their resources. Inventors tend to innovate without considering the timing of their innovation and, too often, create inventions that are too early for the market. By focusing solely on technological innovation and performance, inventors fall into one of several traps, which include:

- Neglecting to consider the manufacturer's costs and capabilities

- Unrealistic views as to the utility and practical use of the invention by the end user

- An absence of a built-in market

Other considerations

Sometimes a company may not take an interest in your invention for reasons that have nothing to do with the merits of your invention or its market position. If a company likes your invention but rejects your offer, it is possible that:

- The engineering department is hostile to outside ideas.
- It lacks the capital to pursue your product.
- A new product makes your invention obsolete.

You may never know why a manufacturer rejects your invention, but if market research indicates the invention has excellent potential, move on to the next potential company on your list.

CASE STUDY: THE RIGHT MANUFACTURER

Sari Crevin, president
BooginHead, LLC
4957 Lakemont Blvd. SE C-4 #293
Bellevue, WA 98007
info@booginhead.com
www.booginhead.com
Phone: (888) 660-6964

BooginHead
Fashionably • Functional

In 2005, I invented SippiGrip®, which is a unique grip material for children that wraps around a sippy cup or a baby bottle. SippiGrip is adjustable, washable, durable, and comes in a variety of patterns and colors. It is designed to attach to high chairs, strollers, car seats, and shopping carts.

I came up with the idea because of my own personal experience as a mother who constantly had to wash cups and bottles that fell on the ground everywhere. Initially, I thought I could find something on the market that would solve this problem, but none of the major stores or online sites carried anything that would help. So, I decided to make one myself.

Starting out, I went beyond my preliminary search to see if any similar products were out there. First, I talked to other parents to see if the SippiGrip was something they would be interested in using and testing. After that, I joined some discussion groups online with parent inventors who shared resources and information with me.

Finding the right manufacturer was a challenge. I was a very small operation that could not afford large minimums when making my product. I needed a company that was willing to work with someone who was starting out, in the process of refining their product, and did not have a lot of capital. I used **www.thomasnet.com** to identify a list of local manufacturers. When I compiled a number of companies, I phoned them and sent homemade samples to get quotes. In most cases, I had the wrong manufacturer, and they kept referring me somewhere else until I found an appropriate partner. At the time, I did not have the understanding or the contacts to know how to license the idea to others.

Eventually, I found a manufacturer willing to produce a low minimum of orders at a reasonable price, yet able to scale much larger if needed. After that, I worked with a graphic designer to put together the packaging, logo, and website. I also gained a lot of knowledge from other inventors about presenting at trade shows to prospective buyers. You really need to stay on top of your supply and re-order, so you are not on backorder for long. Manufacturing in China can have a three-month lag time to process, make, and deliver your orders, which makes tracking your supply and demand levels essential to business. In April 2006 — ten months after inventing SippiGrip — I began selling it online. It was not until December 2006 that I decided to move from treating it as a hobby to really making an effort to have the product distributed in retail stores. As a result, I made the commitment to put more time and money into packaging, graphic design, safety evaluations, and manufacturing overseas. I still work full time, only now I run my own licensed business, BooginHead, LLC, to market the SippiGrip.

My investment to date totals about $30,000. I am just beginning to see financial returns while targeting a profit of around $40K. I filed for provisional patent, and more recently, a nonprovisional utility patent. I was such a novice when it came to this industry and made many mistakes along the way. In retrospect, I would not have rushed to file for a trademark and provisional patent right from the get-go because I lost some money in having to rename and redesign the product. The biggest thing that helped me get SippiGrip to retailers was attending the ABC Kids Expo trade show in Las Vegas. I officially launched my product there, got honest feedback from the retailers, and gained a better sense of what they were looking for. Presentation is everything at these trade shows, so make sure your booth is eye-catching, and you have enough units to show around.

Before You Get Your Patent

You may decide the best way to protect your invention from infringement and the best way to ensure your idea stays in your possession is to obtain a patent. Before you pay thousands of dollars in the hopes of getting a patent, you will need to do the following things.

Patent Searching and Prior Art

To qualify for a patent, your invention must be unique. In order for a patent to be unique, it must be nonobvious technical information and not have any qualifications that bare resemblance to another registered patent. To make sure your invention's technical information, qualifications, and features are unique, you must conduct a patent search for prior art. **Prior art** is anything considered existing information. A patent search for prior art will also help determine if your idea is novel and nonobvious. It also ensures nothing like your claim has been made public before. Prior art includes:

- All previous patents

- Any publicly known, used, or disclosed information in the United States that precedes the inventor's documented date of conception of the invention

- An earlier invention by someone else

- Your invention itself if it was publicly disclosed more than 12 months before the patent application

- Any literature or publication in any language made public before your date of invention or more than 12 months before the patent application, including corporate brochures, trade journals, magazines, academic publications, and any document available in any library anywhere in the world

A patent examiner, who may be an agent or attorney, can orchestrate a patent search. The inventor can orchestrate the search if he or she knows where to look. To be safe, most inventors use a combination of both by conducting their own prior art search and allowing a patent examiner to refine that search using their his or her own legal expertise. The patent search is an effort to determine whether any prior art or evidence of the invention you claim is novel and unique exists in the public domain. Conducting a patent search prior to filing an application is not required, but it makes a lot of sense. Patent applications are difficult, time-consuming, and often costly to file. Your application will be rejected if any prior art matches the claims of your invention. A patent search on prior art will be one of the first steps in developing your market research report.

Provisional patent applications (PPA)

If two or more inventors try to patent the same invention, the USPTO will uphold the application by the inventor with the earliest date of conception. Applying for a provisional utility patent establishes your priority in case someone else submits a regular application on the same claims. Therefore, always look for provisional patents in addition to regular patents when conducting an overall patent search for prior art. If someone's claim on a patent is "pending" by virtue of a provisional patent, you cannot infringe on those claims. To send an application by mail, use Form PTO/SB/16, which may be used as a cover sheet, available at **www.uspto.gov/web/forms/sb0016.pdf**, and send a printed version to:

Commissioner of Patents
P.O. Box 1450
Alexandria, VA 22313-1450

If you use the provisional application process and file a regular application in 12 months, you then have six months before your invention's description is made public. If you file a regular patent application, you have the full 18 months before the competition can learn the particulars. If you significantly improve, change, or redesign your invention in the time between the provisional and regular applications, your provisional application may not have any value if you are going to do a lot more work in the idea in the interim. After the 12-month expiry, all claims on a foreign patent application also will be disqualified. *Foreign patent applications are necessary for expansion into foreign markets and will be discussed later in Chapter 12.*

Patentability assessment

Some claims are patentable, while others are not. However, a claim that is not patentable may still be able to be commercialized and licensed. In other words, you may still commercialize an idea that is considered obvious and not novel; you just do not have any rights to your products claims if someone copies the product. For example, if you build a vacuum cleaner with no unique features other than its basic utility function of sucking dirt into a bag, you will be able to commercialize your product but retain no rights if someone else builds the same vacuum cleaner. On the other hand, if you build a lightweight vacuum cleaner with Root Cyclone™ and Ball™ technology that allows the machine more mobility and greater suction power, you might be violating patent claims the Dyson Company holds. At the end of your patent search, you will know what conditions exist on your claims and what your options are, and the results will influence your intellectual property and commercialization strategies. The first thing that patent examiners will do is scrutinize which of the three types of patents your claims fall under (utility, design, or plant). A utility patent must be considered useful, legal, and not used for deceptive purposes. If the invention has no real-world application, it would not qualify for a utility patent. It may, however, qualify for a design patent. *See the next chapter for the different patent types.*

The next factor a patent examiner will look for is novelty. Claims establish the unique features of your invention. The novelty of your invention does not have to be substantial to qualify. A minor variation on prior art is sufficient to pass a patent examination. Physical differences, new combinations of old materials or components, or new uses for existing items all meet the novelty standard.

The last test a claim must pass with respect to a prior art is the nonobvious nature of the claim. The patent examiner must determine which industry correlates with your invention and assess the degree to which people within that field may have previously had knowledge of your technical information. Courts have established criteria to provide some guidance in testing for nonobviousness. Your claim will be considered nonobvious if your invention involves an original insight or produces unexpected results. If your claims do not meet the first two criteria, there are secondary standards that may qualify your invention as nonobvious. These secondary factors include:

- An invention's ability to resolve a longstanding problem in the field
- An invention that others have failed to make work
- An existing invention modified by omitting one of its primary elements while retaining its effectiveness
- A new feature that enhances the invention's utility
- An invention that accomplishes what had been thought impossible
- An invention that constitutes a commercial success (while not violating the 12-month rule)

In addition to meeting all utility, novelty, and nonobvious standards, the examiner also assesses whether the inventor's patent claims could be made strong enough to withstand competitors' efforts to engineer around the patent. Patents also must be amended continually to reflect innovations and developments that improve your invention. If your intellectual property rights are limited to the point where they do not offer real protection from competition, your ability to license or sell your invention or attract investors will be harmed.

How to conduct a prior art search

Hiring a patent examiner to conduct a prior art search is the safest way to ensure patentability of an invention's primary claims. However, as the inventor, you may be more knowledgeable regarding the technical information you are trying to verify as original; therefore, a prior art search — whether it is the primary search or secondary to an examiner's search — should be conducted. If your invention is simple, or the claims on its technical knowledge are not complicated, it may not be necessary to incur the $500 to $1,200 fee for hiring a patent attorney. However, the more specialized or technical the information is on your claims, the more likely you are to need a professional examiner. This is especially true of niche markets. For example, if you were to invent a new component that enhances the way vacuum cleaners pick up dirt, your patent examiner would be the most qualified to examine your claims against existing patents to determine the possibility for patent infringement. The examiner would look at Dyson's line of patents, including its latest Root Cyclone technology, which consists of a core separator that establishes an extra cyclonic stage between the inner and outer cyclones of its own vacuum cleaner. A patent examiner would be able to examine all the claims and know by his or her experience which specialized types of claims you might or might not potentially infringe upon.

To begin a patent search for prior art, start with online databases such as the ones mentioned in Chapter 1. Before using these databases, select a list of keywords that narrowly define your invention's claims. Use of these databases typically ranges between $50 and $120 per hour, so having preselected keywords will help reduce the amount of time spent searching and cost less. Your first group of keywords should identify patents that may fall in the same industry. When your keywords return your results, they will be associated with a class number and a subclass number. Classes and subclasses are identifiers for registered patents in the Index to the U.S. Patent Classification System, which is the USPTO's patent classification system categorizing all registered patents. The index falls into approximately 400 classes, divided by title description, class number, and subclass number, located at **www.uspto.gov/web/patents/classification/ uspcindex/indextouspc.htm**. When searching the classification system, take the following steps:

Step 1: Using your database, cross reference the class and subclass with the USPTO's Index, and pick the subclasses that are most similar to your claims.

Step 2: Go to the Patent Classification home page at **www.uspto.gov/web/patents/classification**.

Step 3: Enter the subclass numbers. The abstracts derived from the results will give you a fast read on what claims have been registered under the patents you have found. (Another way to reach this page is by going to the USPTO home page, clicking "manuals and guides," scrolling down to "search aids," and clicking "Manual of Patent Classifications.")

If none of the registered claims match your own, move on to the next subclass number on your list.

When it comes to any indexed document that appears similar or too closely related to your invention, first read the document's **abstract**, which is a summary of the invention's properties and claims. Next, look for the classification numbers this invention fits under, as they will provide more places for you to check for related prior art. Also, near the abstract is a section for related documents — publications that contain information relevant to the invention. Many documents also will come with drawings. Look at the drawing to see if it has components that appear similar to your invention. Afterward, search the claims section, which lists the unique features or components of the patented invention. Compare them to see if similar features and components exist for your invention. For any inventor, a prior art search may feel tenuous. If anything, spending time on a patent search for prior art will give you an idea as to whether a patent examiner would be better suited to execute a search.

One last way of searching for prior art is to locate a depository library. A **depository library** is a specialized library, usually located within a university or designated public library, containing information for more than seven million patents. Approximately eight types of libraries have been designated "depository" such as executive departments, service academies, and independent agencies, and there are 1,250 depository libraries. To find one in your area, go to **www.uspto.gov/products/library/ptdl/locations/index.jsp**. Or, if you happen to live in the Washington, D.C., area you can visit the USPTO's Patent Search Facility located at:

Madison East, 1st Floor
600 Dulany St.
Alexandria, VA 22314

Hiring a Patent Examiner

Patent attorneys and patent agents require different fees and differ on the type of services offered. Both, however, are considered patent examiners. In the event you do find prior art, consult a patent examiner before giving up on patenting your invention. Patent law is a highly complex and technical field, and you may have glossed over important evidence that proves your claim is still patentable. If you do hire a patent examiner and want to discuss your claims personally, hire someone in your area. If working with one over the phone is acceptable, look for an examiner who lives in the D.C. area because they will be able to access the Patent Search Facility in Virginia. To search for a list of patent examiners, consult the *Martindale-Hubble* directory, available online at **www.martindale.com** or at your local library. Keep in mind most patent attorneys work only with corporations. Those who offer their services to independent inventors can be found in the yellow pages under patent attorneys. If you want to find a patent attorney that other independent inventors can vouch for, try contacting the United Inventors Association (UIA) at **www.uiausa.org**.

As previously discussed, patent attorneys and patent agents differ in their level of experience, and patent attorneys incur higher fees than patent agents. Whomever you choose, it is your right to question him or her about education and experience regarding intellectual property law and whether he or she belongs to professional associations related to the patent field. Reputable patent attorneys and agents are both registered with the USPTO, so make sure that any attorney or agent you meet is so registered. To become registered, agents and attorneys must have some type of technical background or education and must demonstrate knowledge of the patent application process by passing an exam.

Although registered patent attorneys and agents can conduct prior art searches, patent attorneys can perform additional tasks that patent agents are not qualified to perform, such as legal work involving contracts, trademark, copyright, and handling court cases should litigation over intellectual property result. If you anticipate handling license agreements with manufacturers and distributors on your own and do not anticipate additional legal counseling, hiring a patent agent for a prior art search will suffice.

CASE STUDY: PROFESSIONAL LEGAL ASSISTANCE

Adam Philipp, founder
Aeon Law Group
1218 3rd Avenue, Suite 2100
Seattle, WA 98101
adam@aeonlaw.com
www.aeonlaw.com
Phone: (206) 217-2200
Fax: (206) 217-2201

I am an intellectual property attorney practicing patent procurement, patent portfolio management, intellectual property strategy and planning, intellectual property licensing, and technology law with Aeon Law Group. Aeon is an intellectual property, technology, entertainment, and litigation law firm. I have been involved in the prosecution of patent applications in computer sciences, electrical devices, and related fields since 1998 and with Internet and technology-related law since 1995. I also counsel clients on patent portfolio strategy, patentability, and infringement matters. My area of emphasis covers a wide variety of patent, trademark, and trade secret issues.

Having worked with countless inventors, it is my belief that hiring a patent attorney or a patent agent is critical to an inventor's success. When looking for a legal consultant, do your research. If he or she has not passed the patent bar, he or she is not allowed to file documents on your behalf at the patent office. If he or she does not specialize in patent law, he or she is not the best person to advise you.

Some intellectual property law firms provide free initial consultants or meetings. These, typically, do not include legal advice, but they can include pointers and good practices that apply generally to inventors. Even if inventors pay for one hour of advice, that is often worth it because it can help set the inventor's strategy and provide the inventor with timelines and deadlines.

The most important criteria for selecting a patent attorney is finding somebody the inventor can communicate well with. The patent attorney translates your technical vision into a legally protecting document. If you do not communicate well, or you do not feel comfortable with the partnership, choose another attorney.

Experience is critical in selecting a patent attorney for complex matters. Although law schools offer classes on intellectual property protection, they do not offer much practical experience. The patent legal field is very much a guild system with masters, journeymen, and apprentices. Like any other asset, patents have value. The higher their potential value, the more worthwhile it is for the inventor to invest in an experienced patent attorney to maximize that value.

Getting a patent attorney with the experience to draft a patent application with the right scope is important. Patent attorneys have developed their processes to craft patent applications broad enough to give coverage but narrow enough to not infringe on prior art, and that skill comes with years of experience. Some inventors will need attorneys with a background in the technical field of the invention. For example, some intellectual property attorneys have engineering, software, or science backgrounds that enable them to understand the technical jargon and claims involved with their clients' inventions.

The patent attorney's job is to provide legal advice. They may provide inventors with introductions to people who do industrial design, manufacturing, or finance if need be, but their expensive time is best used for the precise tasks for which they are trained: helping the inventor develop and pursue IP strategies.

Poor use of an IP attorney would be asking him or her to seek licensing deals for you. That is outside the normal area of expertise and very expensive. An IP attorney's role would be involvement in creating license agreements for the inventor or reviewing proposed licenses from the licensee. Patent attorneys, typically, do not involve themselves in commercialization strategy or other business decisions, other than advising on how patent timelines work.

For patent attorneys to be effective, inventors cannot be too intimidated to talk with them. Some clients do not want to look bad in front of their attorney, so they do not ask a question that could lead to greater mutual understanding of the inventor's goals. Some inventors want to avoid communications for fear of being billed. Working out a flat fee for particular services can set an inventor's mind at ease about calling or e-mailing the attorney when a question or new information arises. The patent attorney's job is to protect the inventor, and he or she can only do that effectively if the inventor communicates his or her needs.

A typical scenario involving a patent attorney would be when an inventor wants to submit an idea to a company. It is standard policy that most large companies that they will not agree to any confidentiality or obligation for unsolicited ideas. The inventor increases his or her chances of being considered when they wrap some type of intellectual property around it. A patent attorney can help inventors determine whether they would benefit from a package of IP protections such as a patent and a trademark or a patent and a copyright. In software, it is very common to have copyright and patent working together. In the apparel and outdoor industries, you sometimes have design patents and copyright. Having the advice of a patent attorney can help you create the appropriate IP protection package for your invention.

Understanding Patent Laws

When the search for prior art is complete and you believe your claims stand a reasonable chance of being accepted by the U.S. Patent Office, it will be time to choose what type of patent to apply for, as well as any additional forms of protection, such as trademark or copyright. Applying for the right forms of intellectual property protection are equally as important as making the right claims. Although the USPTO rejects thousands of applications a year because of an inadequate prior art search, choosing the wrong type of patent may inadvertently limit the scope of your protection against infringement. This chapter will discuss what forms of intellectual property protection to choose to maximize your rights, as well as what mistakes to avoid during the application process.

The Three Types of Patents

To profit from your invention, you must be able to control who uses the unique aspects of your invention, so you can charge them for the privilege of that use. The unique aspects of your invention, the parts that require protection and secrecy, may or may not be patentable. Most inventors assume they will require a patent, and many patents are issued every year. Many successful products, though, are not based on patented technology, and most patented inventions are never commercialized into products for sale to consumers or corporations. If you are going to need a patent to protect and commercialize your invention, you will have to decide which type of patent to apply for. There are several different kinds of American patents: utility patents, design patents, and plant patents.

- Utility patents protect the unique functional or useful elements of an invention and will cover any new process, machine, manufactured article, combination of chemicals or materials, improvement on an existing invention, or combination of the five. Some specific utility patent examples include computer software, cable wires, car transmissions, and new drugs, as well as newly developed forms of bacteria and genes.

- Design patents protect the unique aesthetic or design elements of the invention. Some examples of design patents include a new team uniform logo or the spoiler on the back of a car.

- Plant patents protect plants reproduced through cuttings or grafts. The plant must be asexual, and the rights now can be controlled under a utility patent.

All three types of patents require proof that the invention is useful, novel, and nonobvious based on specific descriptive claims submitted in the patent application.

If you wish to safeguard your ideas for as long as possible, it is sometimes to the inventor's advantage to put off submitting a patent application for later in the commercialization process because once the patent application is approved, it will become public knowledge and could be subject to copying. How is this dangerous to your intellectual property? Let us suppose you submit an application, and in the process, you make the wrong claims. Your application will be rejected by the USPTO, and you could lose priority on the claims you are trying to make. During that time, someone could submit the claims you intended to make and establish priority on your idea. By putting off the application for later, you give yourself as much time as possible to figure out how to commercialize while protecting the secrecy of your invention without having a patent. For this reason, a prospective licensee may prefer to jump ahead of the competition by developing a product and positioning it for distribution and marketing before filing a patent application. Rather than assuming it is in your interest to immediately seek maximum legal protection for your ideas, develop an intellectual property protection strategy in conjunction with your commercialization strategy. Inventors should examine the industry trends they discover through market research and use their findings as a guide to developing the strategy for how they will protect their product.

For example, if you are developing a new pharmaceutical drug and deciding how to protect your patent, the *Journal of Pharmacy & BioAllied Sciences* reports that a large number of drugs are losing patent protection and becoming generic, therefore making the pipeline of new patented drugs too small to generate enough profits for future growth. As a result, "the pharmaceutical industry has moved to accelerate the drug development process and to adopt different strategies to extend the lifetime of the patent monopoly to provide the economic incentives and use it for drug discovery and development." In one of the most famous examples regarding the commercial impact of patent expiration, it was reported that Eli Lilly's drug Prozac lost 73 percent of its market share within two weeks after its patent expired and generic drugs were allowed to flood the market. According to the Journal, "As research costs skyrocket, generic drug companies sit poised and are ready to compete [offering a less expensive generic product] as soon as a patent expires." Inventors of new drugs should become aware of this trend through market research and set their commercialization strategy to seek a patent term that fends off generic competition by extending its product life cycle long enough to recoup the return on investment and generate profits.

According to the Journal, corporations are now using commercialization strategies that extend a commercial drug's life cycle before the imminent entry of generic competitors. They are achieving this by patenting the drug's general compound and its uses, then extending its patent protection by "obtaining additional patents covering new formulations of the known compound clinically superior to the previous drug formulation." Once the company extends the length of protection through a new formula, it can continue extending protection by seeking patents on new ways to administer the drug. For example, when the migraine treatment drug Imitrex was set to expire in 2006, the company that sold the product developed and obtained FDA approval and patents directed to Imitrex formulations for intranasal delivery. In the case of the pharmaceutical industry, a drug inventor would use what they learned from industry research to prepare ways of extending product life cycle. They would identify patent extension strategies, such as how they could develop new formulas and ways to administer their drug after their product has been patented and released on the market. In doing so, they will have protected and maximized their product's commercial viability.

Types of Intellectual Property

Different forms of intellectual property are subheaded under these three types of patents. They define exactly what rights will be protected and include provisional patents, copyright, trademark, know-how, and trade secrets. Each of these will be covered individually in the following sections. Every application under a patent type has different features. They include rights conferred, aspects of the invention that are protected, requirements for protection, and length of protection. Normally, inventions will be eligible for more than one type of protection. The stronger an inventor's intellectual property protections, the more leverage the inventor has to license the idea. However, only an invention able to be commercialized will be licensed, no matter how well protected. Also, an invention could be patented, but the claims could be weak and easily circumvented, so the intellectual property protection's strength is vital to the strength of the case made for the invention's protection.

By seeking patent application, you establish priority, or the ability to prove you were the first person to come up with the idea for the invention. Intellectual property protection offers the inventor with priority the right to profit from the invention in commerce; this right is enforceable in civil court. This means that to stop others from infringing on your rights by profiting from your invention without your permission, you have to sue them.

A patent is the strongest form of property protection an inventor can have if it is based on strong claims. A strong patent claim is a claim that is clear, complete, and supported. When making a claim, do not use vague language like, "such as," "when required," or "a major part." These phrases only cause the reader to speculate about the claim and force them to make a subjective judgment rather than an objective observation. The claim should cover all the inventive designs, features, and functions. Describe in complete detail so the reader has a specific understanding of the invention in its context. All claims should be supported by the description. Whatever claims you make have to be fully explained in the description. In some cases, it is possible that only some claims on an application will be rejected, so make as many claims on your invention as possible. Each claim should consist of a sentence that contains an introductory phrase identifying the category of your claim; the body of the claim describing its specific designs, functions and purpose; and the connection between your invention and the category it falls under.

Some of these tools apply to the functions of your invention and others to its design. Some apply to knowledge you keep secret and others to information that is publicly available about your invention. Many of these tools can be used in conjunction with each other to protect an invention, forming a package of intellectual property rights that increases your protection and the value of your invention to a prospective licensee.

The process of intellectual property protection is different for every product. The legal protections available to an inventor are based on the nature of the unique aspects of the invention, how those unique aspects function, and how they relate to other inventions in the same field. It is difficult, therefore, to generalize about the intellectual property protection process. However, intellectual property protection involves:

- Identifying what is unique about your invention
- Researching whether it has ever been invented by anyone else
- Determining what types of intellectual property protections are available to you based on the functions of your invention
- Developing a strategy for integrating pursuit of intellectual property protections and your commercialization strategy

Provisional patents

A provisional patent application is used to establish an early filing date for a patent you plan to file later under a regular patent application. A provisional patent is useful in the early stages when an inventor is trying to determine the viability of his or her idea in the marketplace. It contains a description of the invention but does not include the legal claims required for a regular patent application. Provisional patents help an inventor commercialize an idea because it is much less expensive than filing a Regular Patent Application, and it proves he or she had the idea before anyone else based on the early filing date. Submitting a provisional patent application permits the inventor to claim patent pending for his or her invention even if the inventor has no plans to pursue a patent. *You will learn more about filing provisional patents later in this chapter.*

Trade secrets

A trade secret is any confidential information that must be protected against competitors. Trade secret laws under the Uniform Trade Secrets Act protect your confidential information, and in this instance, no formal patent registration is necessary. As long as your trade secret is kept under wraps by nondisclosure forms and avoidance of public disclosure, the trade secret is protected perpetually unless someone else reverse engineers the ingredients or independently discovers the same process, set of ingredients, or combination of the two. The Uniform Trade Secrets Act can be enforced for your information if you can prove the information provides a competitive edge; measures to maintain secrecy were used; and a competitor acquired your secret information through improper means such as espionage, bribery, breach of contract, or misrepresentation. Under most state laws, you forfeit a trade secret if you willingly divulge the trade secret without a confidentiality agreement or make the information public.

Know-how

Know-how is the inventor's expertise regarding the invention that facilitates implementation. Know-how can provide a basis for a licensee or purchaser of an invention to pay the inventor as a consultant to provide intellectual services. Know-how and trade secrets generally are not patentable but may aid in the description of what property rights are being protected under law.

Trademarks

A trademark is a brand name, word, or logo that represents the product being brought to the public. To be protected, the mark must either be used in commerce or registered with the intent to use it. Although commercial uses are sufficient to establish trademark rights, registration with the USPTO can strengthen trademark enforcement efforts. The TM symbol next to a word, brand, or logo is sufficient trademark designation and is the designation for all nonregistered trademarks. A trademark that has been registered with the USPTO is designated with the ® symbol. Use of the ® for a nonregistered trademark could interfere with the right of an inventor to subsequently register the mark. Trademarks can also be registered with a state attorney general or with an industry trade.

Some trademarks identify products or brands; others identify services including service marks, certification marks, and association marks. Trademarks that identify companies are known as trade names. Legal protection extends to distinctive and recognizable packaging that consumers automatically associate with a particular brand. Using distinctive packaging colors, shapes, and ornaments on packaging, uniforms, buildings, trucks, and other objects is known as trade dress. *Trademarks will be covered later in this chapter.*

Copyright

Copyright protects original written works, artwork, music, and computer software. Copyright protection takes effect as soon as a work is made public through publishing or performance. Although publication automatically establishes copyright, enforcement can be strengthened by registration with the Library of Congress. To protect an original written work, register online at the U.S. Copyright office at **www.copyright.gov/eco**.

Copyright law protects "works of authorship," which include literary works such as short fiction, short stories, novels, nonfiction articles, poetry, newspaper articles, newspapers, magazine articles, magazines, computer software, software manuals, text advertisements, manuals, catalogs, brochures, and compilations of information, such as databases. Publication has a technical connotation in copyright law. According to U.S. Copyright statute, "Publication is the distribution of copies or phonorecords of a work to the public by sale or other transfer of ownership, or by rental, lease, or lending. The offering to distribute copies or phonorecords to a group of persons for purposes of further distribution, public performance, or public display constitutes publication. A public performance or display of a work does not of itself constitute publication."

Why should you register your work with the United States Copyright Office? There are several reasons. For one, to sue someone for copyright infringement, the owner of the work must first register the work with the U.S. Copyright Office.

You may register the work after someone has infringed upon the work, but the registration will only be relevant to infringements that occur after the registration. On the other hand, if you register your work within 90 days of publication, the statutory damages provisions apply to infringements before and after the tangible registration. Registered works may be eligible for statutory damages up to $100,000 and attorney's fees in successful litigation. If the registration is completed within five years from the creation of the work, it is considered prima facie proof in a court of law (in other words, proof given of something's existence on first appearance). Prima facie proof provides evidence that the existence of the invention was noted in the court of law on a specific date and strengthens protection against infringement. Registration is inexpensive, about $20 per work registered, and relatively straightforward. To register, the author fills out the copyright application and mails it to the U.S. Copyright office with a check and a nonreturnable copy of the work (one copy if the work is unpublished, two copies if it has been published). Published works that owners want copyrighted have to be registered within three months of the publication. This is called "mandatory deposit."

The United States Copyright Office has issued a Notice of Proposed Rulemaking, which will modify the way it accepts group registrations of individual works. The proposed change would affect writers of short works, which they group and register collectively; it also would require electronic registration. Please note that the copywright office will not accept delivery by means of overnight delivery services such as Federal Express, United Parcel Service, or DHL.

Copyright law broadly defines pictorial, graphic, and sculptural works as including all two-dimensional and three-dimensional works of art (fine art, graphic art, and applied art), photographs, prints, reproductions, maps, globes, charts, diagrams, models, and technical drawings, including architectural plans. This can encompass everything from sculptures and paintings to less conventional items, such as mannequins and decorative belt buckles. And as with other works of art, the required level of creativity is minimal, so this includes everything from realistic photographs to drawings and renditions of a product.

The emergence of the Internet has made the infringement of copyright works simpler and harder to hunt down. Digital technology makes illegal copying or piracy far easier, whether the piracy involves a major film, recorded song, software package, electronically published report, or personal travel photo. Originators should seek professional advice on protecting their work. Some experts feel the intellectual property situation on the Internet is getting to its worst point ever, as this area requires knowledge of technology and law. When an author places his or her work on a home page, it can be viewed by people all around the world, even in countries that have no copyright treaty with the United States. If the author's work is infringed upon, he or she may never even be informed about it if the work is reproduced in a foreign country. In addition, it is costly to pursue a copyright violation lawsuit in a foreign country.

When to Pursue a Patent

Patent applications are expensive and complicated. The average cost of getting a utility patent, including patent attorney fees, is between $10,000 and $15,000. Inventors will know when it is worth their time to pursue a patent depending on two factors: the patent's profit-enhancing capability and its estimated life cycle. Life cycle is the amount of time a product is able to retain market share in the face of competitor reactions and the entry of a new generation of products or new technologies into the market. A patent can create higher profits by maintaining a monopoly on some intellectual property claim that increases profitability. The longer the estimated life cycle of a product, the higher the expected profits due to its staying power on the market. Other factors pertaining to profitability include any decreased cost of production or liability. Even if a patent can increase profits, will it increase profits enough to exceed the cost of obtaining and maintaining the patent?

The total costs involved in bringing your invention to market must be projected before you (or the licensee) can expect to profit from an invention. Without patent protection, how long might it take for the person creating the product to see a return on their investment? If that time is longer than the life cycle of the product, a patent may not be worth the investment. In addition to these considerations, the inventor should consider whether the design will be the final version. A design patent on an interim prototype subject to design change is unlikely to offer protection if altered and would not have much value you or to a licensee. When all considerations have been made, pursue your patent claims only when you are certain that you have finalized your invention's design.

The Scope of Intellectual Property Protection

In the United States, patent protection lasts for 17 to 20 years. A patent gives the inventor the right to prevent others from making, selling, or even using a product based on the claims made in the invention's patent. Patents are personal property that can be sold or licensed by the inventor.

When your invention is proprietary, which means it is owned by an individual, it is your property that you own and control. Intellectual property protections are methods for proving or legally confirming ownership of the ideas, words, and concepts that make up an invention.

Proprietary ownership of your invention gives you the right to use an invention without compensating others. It also prevents others from using it without compensating the owner. A patent does not assure the profitability of an invention, though not having patent protection may prevent you from profiting from your invention. Intellectual property protection is a legal tool. The greater your knowledge regarding the technical aspects of patent law, the more favorable your position will be with respect to earning royalties from sales.

To benefit from patent law, you must develop a patent strategy that works in concert with your commercialization strategy. For instance, it may not always be the best solution to file a patent before doing market research or approaching prospective buyers. The best way to formulate a patent strategy that works in concert with your commercialization efforts is to determine when a patent application should be filed and the ways that protection can be broadened and maximized. The first step is to rank its importance in your commercialization strategy. If the claims included in the patent application that are being protected are limited in scope or do not assist in the overall commercialization process, filing for an application can probably wait. If having a patent registered is central to getting a licensing deal, or if the claims are specific and potentially subject to development by competing entities, file the application first. The next step after ranking its importance is to list your intellectual property assets and determine whom they should be packaged to in an application. To define what assets can be packaged, return to the documenting materials you kept during the experimental phase, and break down the details of the process, design, and functionality required to create the final version of your invention. These details will become the basis for any utility patent claims, trade secrets, trademarks, copyright, and/or know-how.

What features and functions qualify your invention as a utility? What logos, names, graphics, or brand identifiers developed in an association with your product qualify as trademarks? If your invention includes written materials you do not want copied or published elsewhere without your expressed consent (i.e. handbooks, manuals, marketing materials, or articles about your ideas), these materials can be copyrighted. Lastly, what specialized knowledge do you possess for making, manufacturing, distributing, or marketing your invention? This is the basis for protecting know-how, and it protects your right to earn money as a consultant to licensees.

Formal Property Protections

An invention may be eligible for more than one type of protection. The stronger an inventor's intellectual property protections, the more there is for the inventor to license. Patent laws are considered tools that offer legal protection of intellectual property. Intellectual property protection is enforceable in civil court and grants the inventor the right to profit from the invention. To prevent others from infringing on your rights without permission, you have to sue them. The types of remedies a court may apply if a person or organization is found to be infringing on your patent rights include monetary damages and what is known as **injunctive relief**, which means a court order to cease the infringement and perhaps a monetary penalty will be applied if infringement continues.

What claims you file and how you file them will determine both the architecture and enforceability of property protection. Some tools under patent law will apply to the functions of your invention and others to its design. Some forms of protection can be used in conjunction with others to strengthen the overall level of protection, forming a package of intellectual property rights that increases the value of your invention to a prospective licensee.

Provisional patents continued

As discussed earlier, a provisional patent filing is an early patent application and a record of your intention to file with the USPTO later. The provisional patent application contains a description of the invention but does not include the legal claims required for a regular patent application. Submission permits the inventor to claim a patent pending for his or her invention even if the inventor has no plans to pursue a patent. Filing a provisional patent early in the commercialization process makes sense whether filing for full rights ranks high or low on your list of priorities. A provisional patent application must contain:

- A description of the invention
- Sketches or drawings that demonstrate the process (if applicable)
- Cover sheet and transmittal form
- Check or money order
- Return receipt

Remember, a provisional patent application is not a regular patent application; therefore it does not require the submission found in a regular application, such as a patent application declaration, an information disclosure statement, patent claims, an abstract or summary, a description of the inventor's background, or a description of the invention's advantages. What a provisional patent does provide is that you will be considered to have reduced your invention to practice, even if you have not built a prototype or tested your claims. Claiming **reduction to practice** means you can claim the application date to overrule your opposition's prior art (if later dated) and establish priority. The purpose of building a prototype — in addition to having a model to show manufacturers — is that it demonstrates the reduction to practice that is essential to filing a regular patent. With respect to patent law, filing a provisional patent eliminates the need for a prototype, as well as the costs associated with building one. If you decide to file a provisional patent, make certain you have filed correctly, otherwise you may not be able to rely on the filing date as either proof of reduction to practice or proof of prior art. If you plan to file a regular patent, make sure you do so before your provision grant expires (12 months from the date of application). If you fail to apply for a regular patent and the provision expires, someone else has the right to file a provision or regular applications on the same claims.

Regular patent applications (RPA)

To receive a regular patent, an inventor must have reduced his or her invention to practice through documentation of experiments or assembling a prototype. The patent application form will require you to provide:

- A written summary or abstract describing your invention and its purpose

- A graphic representation of your invention and its parts (a drawing or sketch)

- A specific field under which your invention falls (the USPTO has a list of invention categories in its Manual of Classifications)

- Some background or history on the category of your invention, which states the need for your invention and how it improves on the status quo

- A narrative description of the drawing

- A detailed narrative description of all the materials, components, processes, and sequence of events ideally involved in using your invention

- A series of claims that will establish exactly what the limits of your protection are

The two main sections of the application are **specifications**, which cover what the invention is made of and how it works, and **claims**, which are examined in court if litigation against infringement is pursued. Both of these components are required for your patent application to be approved. The specifications section should provide enough information for a qualified expert in your field to understand and be able to apply in practice. The section should also describe the intended way of using the invention. The more descriptive you are in this area, the more likely you are to have your application approved. However, the more detail you disclose, the more you inform potential competitors how to design around your claims and obtain the same results without paying you a royalty. The claims section defines the intellectual rights you intend to reserve. Your claims are a package that defines the scope of your rights. If your patent examiner discovers any prior art that appears similar, you must file an Information Disclosure Statement (IDS Form 10-5) and list those references in order to have them considered acknowledged by the examiner. By filing this statement, the USPTO notes that you believe, despite the similarity, your claims are patentable over these references, which eliminates any contest the patent owner of those references can make against you later in the court of law.

The Application Process

When you mail a PPA or RPA to the U.S. Patent & Trademark Office, it is recommended that the application be sent via Express Mail, and you should retain a photocopy. The USPTO states that mailing any materials via Express Mail with the EM number on the transmittal letter filled out on your application automatically considers your application as "status pending" as soon as the clerk hands you a mailing receipt. Use the number on the receipt to track the application's delivery. When it has reached the USPTO, allow about four weeks for a response. At that time, you should receive the return receipt from your application as evidence that the PTO has received your application.

Included on the receipt will be information that includes the examining division assigned to review your application. This receipt is an acknowledgment that the USPTO has created a file on your application and begun the process of evaluating it. Between six and 24 months later, you will receive a first Office Action Communication (OAC) letter from the examiner reviewing your application. The OAC will show that your application has been either accepted or rejected. If your patent is granted, you must pay a fee for your patent to be issued. You will be notified of the data upon which your patent will be issued and, shortly after that date, you should receive a letter or deed confirming your patent from the USPTO.

On the return receipt will be an eight-digit serial number with a bar code sticker, which is the number assigned to your application. If the application is incomplete, the USPTO will send a letter stating the deficiency. If you receive this letter, call the USPTO immediately to resolve the issue, and

122 How to Get Your Amazing Invention on Store Shelves

have your serial number on hand. If any joint-inventors are involved, they are advised to complete a Joint Owners Agreement (Form 16-2), which should not be mailed to the USPTO but stored as documentation of joint ownership of the invention. In the event that prior art is noted, you may also be required to file electronically using Form SB/28, found online at **www.uspto.gov/web/forms/sb0028_fill.pdf**. If you prefer to file a patent electronically instead of by mail, check **www.uspto.gov/ebc/index.html**. Filing electronically means you will be able to track your application status on the Internet. Be aware, however, that the electronic application process may be more time-consuming. To file electronically, go to **www.uspto.gov/ebc/index.html,** and click on EFS Web Unregistered eFilers. Fill out your name, e-mail, and the type of application being filed and click Continue. New registrants are given a Digital Certificate and password to authenticate their identity each time they log in. To create a new application, click Authenticate and select New Application. This clicks into the Application Data page, which should be filled out and requires the title of the invention, the docket number for the application being chosen, inventor name, and address. To attach supporting electronic materials such as drawings or designs, find Files to Be Submitted, click Browse, and search for the attaching file. When all attachments are selected, click Upload & Validate. When all the information is complete and all attachments are made, the process will bring you to a Review Documents Page. Review all the information being submitted to make sure it is correct, and click Continue. The next page should be the Calculate Fees page, which will ask you to identify yourself as a Small Entity. Your fee as a Small Entity will be calculated and result in a page that lists all files being submitted, as well as the filing fee. After you click Submit, you have completed your application online. For help regarding the electronic filing process, the USPTO's Electronic Business Center can be reached at (866) 217-9197.

If a patent is granted, maintenance fees are required at three and a half, seven and a half, and 11 ½ years after a patent is issued. You will be informed by the USPTO when an upcoming maintenance fee is scheduled to keep your patent current. USPTO maintenance fees are minimum fees, subject to change, and apply to individuals and small organizations. They do not include fees for amendments, late fees, surcharges for excessive number of claims, and appeals. The costs do not reflect any costs of hiring experts to research, write, or review your applications.

Filing an application for a design patent is slightly different from filing a utility patent in that the description in the specifications section will not cover how the invention works, but rather what the invention looks like. A design patent requires filling out a Transmittal Form 10-11, a 10-3 Form with a check or money order, and a specification form 10-10, which states:

- The design's intended use and purpose
- What applications the design is related to
- Federal sponsorship if funded by the government
- Drawing and sketches
- Claims
- Return receipt
- Application Data Sheet (optional)

If you submit an application that is rejected, the OAC will provide a list of reasons, citing anything from prior art discovered or proof of information that renders your invention obvious. You will have a period of several months to respond to the examiner's findings. Allow several months from that point to receive a response from the examiner with a final decision. In the event that an application is rejected by the USPTO, it is possible to file a continuation application. A **continuation application** can be filed if the patent examiner suggests that you try to file a patent under new claims, or there is reason to believe the application was not filed accurately. Class and subclass of any claims are determined by the application's subject matter. Therefore, any patent filed under new claims will be assigned to a different examiner division because the claims will be filed under a different class and subclass. Keep in mind that most patent examiners suggest that an inventor file no more than two continuations and one request for continuing examination, otherwise a court may find the applicant guilty of intentionally delaying the process.

Preparing a continuation involves following the same process as the original application, but with the exception of rewriting the claims on a Patent Application Declaration (PAD) Form 10-1 and resubmitting an IDS Form 10-6. To avoid having to fill out all the paperwork of the original application, you may file a Request for Continuing Application (RCE). An RCE has the same effect as a continuing application, except the form used is a 14-1. The filing fee costs less than having to file a new application and remains fixed regardless of the number of claims in the application.

The One-Year Rule

An invention can be patented by anyone within 12 months of its becoming public if other intellectual property protections have not been established. Anyone given full disclosure to the particulars of your invention should sign a confidentiality or nondisclosure agreement. Witnesses to your experiments, an invention evaluator, a patent examiner, an industrial designer or consultant, and potential licensees should sign nondisclosure agreements. No one beyond this group should have any information regarding your invention.

If you discuss your invention with a potential licensee without having them sign both a confidentiality agreement and a statement saying they understand your discussion is not an offer for sale, it can be considered the start date on the one-year rule if that person documents the date upon which you disclosed the information. As an inventor, you will have to decide how to protect your invention based on how much time and money you are willing to spend protecting it, the value of various protections with regard to licensing your invention, and how various tools mesh with your commercialization strategy.

Finding a Patent Agent

Patent agent services come at a lower cost than hiring an attorney, but they cannot represent a client in court. If you prefer to use a patent agent, look for one in specific jurisdictions at **www.uspto.gov**, and choose one with a background that matches the industry in which the licensing agreement applies. Industries that are likely to require dealing with agents include toys, giftware, and home furnishings. Industries in more technical fields, such as electronic and mechanical devices, are unlikely to require dealing with agents. Patent agent agreements should include a nondisclosure agreement with an agreement stipulating the nature of the agent/inventor relationship. The agreement should also disclose pricing agreements and establish an application filing timeline. It also will be stated in writing that, in the event that if any matter concerning the patent goes to trial or requires litigation, the inventor will need to hire a separate attorney for those services.

Under a typical agent representation agreement, expect to pay the agent a percentage of licensing fees. Provide a prototype or model of your invention, documentation, and marketing materials. Read the agent's agreement carefully, and know what you are agreeing to. It is not uncommon for an agent to express his or her right to profit from any licensing deal during the time her or she represent you, even if another agent you hire brokers a deal. If you are unwilling to accept this provision, find an agent whose agreement does not contain such language.

Legitimate agents travel at their own expense to meet with licensees. They do not accept upfront fees, and they do not represent competing products. If there is a possible conflict of interest between an existing product they represent and your invention, they will discuss it with you. Agents keep inventors updated about contacts they have made and the outcomes of those contacts. Agents are free to conduct their work in the manner they deem best. They have a financial interest in your success.

When considering agents, evaluate their legitimacy, and weed out any who operate like invention marketing scammers. *See Chapter 10 for more information.* How many clients do they work with? What percentage receive licensing agreements? Do they specialize in the industry your invention is related to?

Patent Attorneys

Patent attorneys should draft patent claims in order to provide the inventor the full coverage available by law. Weakly constructed claims may lead to an approved application but may be circumvented by competitors who explain loopholes that allow them to design a similar product without infringing on the claims of the original owner. If that happens, you may have spent thousands of dollars and hundreds of hours without having gained any significant intellectual property protection. For a list of patent attorneys and agents registered with the USPTO go to **https://oedci.uspto.gov/OEDCI/query.jsp**. The USPTO does not require that an applicant use an attorney; however, most inventors use patent attorneys to fill out and file patent applications for them.

Patent attorneys charge the same rates as other attorneys — anywhere from $100 to $400 per hour. Do not use money as the basis for finding a patent examiner. It is better to spend more money per hour and get someone experienced than to engage an attorney who may file an easily rejected or circumvented application; he or she will charge you for time regardless.

The best way to navigate this process is to interview several patent attorneys and select the one you feel most comfortable with (both personally and professionally) and who has been the most clear in communicating what to expect, including what services he or she will provide. Criteria for qualification may also include questions about education (he or she should be an intellectual property law specialist), registration with the USPTO, the number of patent applications filed and obtained, experience with other aspects of intellectual property law, including other protections (copyright, trademark), and other aspects of inventing (licensing, patent infringement actions).

Although you may have little control over the cost of filing and the hourly rate of a good patent attorney, you do have some control over the amount of billable hours required. The most complex aspects of the patent application are searching for prior art and drafting claims. If an attorney is given a list of prior art results to work with before he or she begins a patent search for prior art, this will cut down the amount of time spent trying to find any prior art that must be acknowledged on the application forms. You can also limit your costs by hiring an attorney to review your application after having done most of the groundwork. Because most of the questions have been answered, the number of billable hours will be less, and you will ensure everything has been thoroughly researched before you submit to the USPTO.

It is the patent attorney's job to help define the patent strategy, as well as decisions about how to use it. For example, an inventor might want to apply an invention horizontally across the marketplace or with modifications enhanced specifically to apply to a vertical market. It is the inventor's job to decide which market is more attractive if the patent cannot cover both markets. The patent attorney acts as adviser and facilitator in crafting the patent based on the inventor's decision about which direction to go.

Patent Deadlines

The patent application process is a deadline-driven process. The first deadline is the 12-month window between when an idea enters the public domain and when a patent application must be filed. Your idea enters the public domain when:

- You tell someone about your invention without restricting his or her ability to share the information through a nondisclosure agreement

- You disclose information about your invention with the intention of selling or licensing the invention

- You or anyone else publishes a document with information about the invention, including a provisional patent application, brochure, academic paper, or magazine article

From the date you have filed a patent application, you will be given a three-month deadline to file an information disclosure revealing any prior art that your searching has identified. You have a legal obligation to turn over any prior art that may influence the patent division's ruling. After you have filed a patent application, you have one year to file a foreign patent application, otherwise someone in a foreign market has the legal right to design, manufacture, and sell a product that infringes on your original claims.

The USPTO will publish your patent application within 18 months of submittal. It may be another 18 months before you receive a patent if one is awarded. There are a variety of drawbacks to having your patent application published. In the time between publication and patent award, a competitor can read your application and start trying to circumvent your claims to capture the market with a competing product before you are able to manufacture your invention. Publication also notifies any parties identified with prior art or competitors who do not want your idea patented to challenge your application by submitting what they believe is their prior art or grounds for rejection. If your application includes trade secrets, they will no longer be secret, and you lose all rights to protection.

Patent applicants can request that the USPTO not publish their application through a Non-Publication Request (NPR). That way if your patent claims are infringed upon during the period when your application is public and pending, you can still claim damages against an infringer. Having an NPR established does restrict you from filing a foreign patent application. If you file a foreign patent application while you have an NPR in place, your U.S. patent application will be ruled abandoned. You can request your NPR be revoked, if you have one in place, and later decide to file a foreign application. Failing to revoke NPR may have serious legal consequences, such as litigation, so be sure to revoke it when required.

Trademark Protection

A trademark symbol is a special design or word used to represent a service or product. The Pillsbury® Doughboy™ is a famous registered trademark. Building designs can also serve as trademarks or service marks. For example, the McDonald's Arches are a famous design registered to McDonald's Corporation for restaurant services. Other registered trademarks include the Puma® Company's design and the Ford emblem. Pictures or drawings of a scene or character, such as the Pink Panther (a registered trademark of the Corning Company) or MSN's Butterfly (a registered trademark of Microsoft), are often used as trademarks or service marks. The Apple® iPod is a registered product shape, a form of trademark.

A product or container shape can also function as identification and therefore, can be an enforceable trademark. Can a sound be trademarked? Yes, the three-tone chime of NBC has been registered as a service mark. Sound trademarks were in the news recently when Harley-Davidson® declared that it was attempting to register the exhaust sound of a Harley-Davidson motorcycle. A trademark also might be a combination of letters and a design, such as IBM or Silicon Graphics. Logos are almost certainly the next most common form of mark. A logo can be depicted as a design that becomes a mark when used in close relationship with the goods or services being marketed. The logo mark does not have to be complicated; it only needs to differentiate goods and services sold under the mark from other goods and services. Dunkin' Donuts'® trademarked label is an example of this. Phrases or taglines can also be converted into trademarks:

- *"Don't leave home without it."* — American Express
- *"The greatest show on earth"* — Ringling Bros. Barnum & Bailey Circus
- *"Everywhere you want to be"* — Visa
- *"Just do it"* — Nike

Trademark protection must either be used in trade (a product offered for sale) or registered with the intent to use it. Although commercial intent is sufficient to establish trademark rights, registration with the USPTO can strengthen trademark enforcement efforts. Because there is no evaluation for a trademark other than its attachment to commercial use, a trademark owner cannot be sure how a trademark strengthens an owner's claim to intellectual property rights until the trademark is introduced in court. The strength of an owner's claim to a trademark is distinctiveness. If you or someone else has devised a unique and nongeneric logo that does not resemble the logo for any other known product, then it is considered distinctive.

The more generic the name of your product, brand, or company, the less distinctive it will be considered. In this case, your trademark is not likely to strengthen the claim on your patent. Trademarks can strengthen the patents on products with less generic names. For example, attempting to register a trademark for "Curly Fries" would not strengthen the overall patent protection, even if the claims on your curly fries show they are made or designed differently from others. Patent laws also state that a generic term cannot be registered as a trademark. For example, you cannot trademark the name "lipstick" for a type of lipstick you plan to sell. A trademark also cannot use deception, include offensive language or pornography, or replicate the insignia of another party. Nor can it include your last name, a person's image without their permission, or an insult to certain people or organizations.

Trademark protection increases if the trademark starts acquiring wider recognition in association with your product. For example, a cow on a milk brand may have little trademark protection when it first hits the market, but it may gain protection over time if a large enough public begins to recognize and associate the trademark with the product. Trademarks initially entitled to protection can also lose protection if their owners use them as generic terms. Band-Aid® and Xerox®, for example, are marks that have taken on a generic meaning in public parlance.

An inventor can register a trademark with the USPTO before use in commerce, thereby establishing priority for the mark if the inventor plans to use the trademark in commerce relatively soon. Initial registration is good for six months and can be subsequently extended (for a fee) for up to three years. If the trademark is not used before the intent to use registration expires, the trademark is considered abandoned (and it has now been made public so others may adopt it). Trademarks also can be used to address the issue of unfair competition. Unfair competition is using an established product's packaging or other non-patentable design features or advertising.

If a brand of nail polish is marketed in a distinctive bottle, for example, and a competing brand imitates the bottle, even if the bottle is not eligible for trademark or design patent protection, it may be protected under unfair competition law, especially if the bottle shape has become synonymous in consumers' minds with the original product.

Prior trademark searching

The best way to ensure that a trademark has distinctiveness is to conduct a trademark search. By searching, you eliminate the possibility that the trademark designed for your product is not unique. Many trademarks are not registered, which means there is no central database to search that will help you reach an authoritative conclusion that your trademark does or does not infringe upon another mark. There are, however, databases you can scour, including:

- The USPTO registry
- The Secretary of State's office database
- Industry Trade Association databases

Because these resources are incomplete, let us begin with the most complete resource available. If you are using a fictional word, it should be simple to determine whether that term is in use through an Internet engine search on Yahoo! or Google. The USPTO offers "intent to use" registration of trademarks that registrants plan to use within six months. Intent to use registration allows individuals or companies to establish priority, in effect reserving a trademark even though it is not yet in use. The USPTO database will provide access to these trademarks. If you are likely to do business primarily in one geographic region, search the Secretaries of State's databases in the states where you will be conducting business. If someone is using a similar trademark, you may still be able to use the trademark if your product or business is unlikely to be confused with the original holder of the trademark. When in doubt, consult a patent attorney.

Registering your trademark

About a quarter of all trademarks used in business are actually registered. Although your trademark acquires legal protection through commerce, there are additional benefits to registration. The benefits of registration include:

- Providing patent attorneys and judges one more tool for enforcing protection of your trademark

- Making it easy for others who want to use a similar or identical trademark to find that it is being used

- The ability to prohibit infringing Internet domain names

- Giving judges the ability to levy additional monetary damages for trademark infringement

- Enabling judges to restrain infringing parties and prevent them from using the trademark while you await a ruling on your rights

- Stopping the importation of foreign items that infringe on your mark

Once the trademark is registered, you must use a trademark notice phrase (registered with the USPTO) or the ®. When you apply to register your trademark, it is important to specify what your trademark will be used for, as well as the industrial classifications your product falls into. You will be charged a fee for every category you seek registration under. A trademark may have several elements, a combination of words, punctuations, colors, and images. Each element should be examined independently to determine which is the strongest. If a word is the trademark, it is called **typed format**. The way the word is written (the script or the combination of upper and lower case letters) is called a **stylized** or **design format**. Retaining an intellectual property attorney to review and even register a trademark application is advisable, particularly if the application is rejected. Once an application is submitted, the approval process will take about 18 months.

An application may be rejected if it appears to be confused with an existing trademark and lacks distinctiveness. If this should this occur, request that the trademark be placed on what the USPTO calls the Supplemental Register. The Supplemental Register is a list that protects your priority for the trademark while you use it in trade and hope that it acquires distinctiveness through association with your product in consumers' minds over time. Placing a trademark on the Supplemental Register, however, is tantamount to acknowledging that the mark is not protectable. In other words, you may retain a chronological priority with the USPTO, but you will not be able to make a sufficient case in court that your trademark has been infringed.

When your mark is first approved, it is granted preliminary approval and enters a period in which others can challenge or object to your trademark's registration. If someone believes your trademark is offensive or infringing on his or hers, he or she can file an objection that triggers an appeal to be heard before a trademark board. To pursue registration, you will have to hire an attorney to argue your case. If the trademark is refused registration, you can appeal the examiner's decision with the assistance of an attorney.

Although trademark protection lasts for as long as it is in use, USPTO registration lasts for six years. To extend the registration period, apply for an extension by providing evidence and a sworn statement that the trademark is still active. After the first renewal, the trademark will remain registered for ten years and must be renewed every ten years thereafter.

Creating a professional marketing package

Trademarks are important step in the branding of your product. Through branding, you establish consumer recognition for it. Branding leverages the value of recognizable symbols that the market associates with your product, and it must visually separate your professional marketing package from other inventors who solicit business to prospective buyers. Catalogs and potential licensees want to know that the business backing your product has a professional mien and can leverage recognizable symbols that people can associate with things they want in life. To create a professional marketing package, you need to develop four important aspects of branding: a company name, a company logo, business cards, and letterhead. These materials will be used for sending out queries or marketing materials to prospective catalog houses and licensees. Many business owners hire ad agencies that assign account managers to help businesses develop logos, but the cost usually runs between $2,000 and $10,000 for design and development. If you choose not to hire an ad agency, it is important to understand how to create a logo. Logo Creator is a software-based tool available for purchase on CNET at **www.download.cnet.com**. For $30, Logo Creator's features are highly customizable and allow users to create unique, professional looking logos. A **logo** is an illustrative, graphical, or textual representation of your business or product's essence. According to design and programming news blog Webmosiacs.com, there are ten key components of a professional and effective logo, which include its:

- Longevity
- Clarity
- Attraction for consumers
- Ability to express the correct idea
- Intelligibility
- Visibility
- Simplicity
- Retention in the consumer's memory

- Descriptiveness
- Colorfulness

After creating a logo, include it on the letterhead in the marketing materials used to solicit business from catalogs and potential licensees. Put the letterhead in your printer, and write a cover letter to the potential buyer asking for their product submission guidelines or vendor information kit. Also, provide some background on yourself and your business. Do not talk about your day job; present yourself here as a professional product developer (which you are). Buy some nice pocket folders with diagonal cuts on the pocket for a business card along with envelopes at your local office supply store. Consider purchasing large labels, so you can print the prospective customer's name to give your package a polished look. If you plan to include prototypes, obtain a mailing package of adequate size for your prototype, and assemble a few dozen packets. You might, for example, put press clippings in the left pocket and the sell sheet, price list, and terms and conditions in the right pocket. Put the sell sheet on top and business card in the cardholder on the pocket. Do not use paperclips because they tend to form impressions into the paper. Assemble 25 or so packets at a time, and identify good catalog house or manufacturing prospects. If you would like to increase the likelihood of a response to your mailing, create self-addressed, stamped postcards inviting feedback on your product. As you identify catalogs that are good fits for your invention, compose a professional-sounding cover letter, print and sign it, and place it on top of your pre-assembled marketing kit folder. Put the whole thing in a manila envelope with a mailing label.

CASE STUDY:
THE IMPORTANCE OF
ACCURATE PATENT FILING

Jennifer Pines, president
EmmaLu Designs
jennifer@emmaludesigns.com
www.emmaludesigns.com
Phone: (678) 613-3057

When my daughter was young, she kept spilling milk formula on my car seats, and I would have to get the car professionally cleaned. Taking the car to a cleaning service always required removing the car seat cover from its straps, which I found to be a very difficult and annoying task. This happened so frequently that I thought there had to be something that could allow me to easily remove a seat cover when needed. As it turned out, there was nothing available on the market, so I decided to make one myself.

Starting out, I read books on inventing, marketing, mom inventors, business startups, Web business, and design books. I put in 40-hour weeks getting my invention to market and patented while investing about $40,000 of my own money over the first two years. Right now, my business has reached break even, but it is not yet profiting. Because I manufacture in the United States, my cost per unit is very high. I wholesale for very little to get the product out there with a low retail cost. I reasoned that breaking even during the short term was important to making a large profit over the longer term.

I probably was too invested to turn the invention over to a licensee. I worked hard to get it started, and I never thought about letting it go. In retrospect, I was probably paranoid that someone was going to take my idea, so I hired a patent agent and applied for a design patent right away. One day, I was speaking at an inventors association meeting and had been checking on the status of my application periodically. While doing so, I discovered that my patent agent had made a critical error in filing the application, which might have caused the application to be rejected. When I informed my patent agent of his mistake, he was embarrassed for missing something that a novice was able to catch. He immediately refiled the paperwork, no fee charged, and we were back on course. This mishap could have been much worse if I had not been tracking the application to protect my intellectual property. The lesson I learned was an important one: Although you should hire a professional to handle the technical aspects of patent application, always be involved in the process by tracking your application and making sure your patent attorney has explained the process so you understand it.

Licensing to Companies

Without intellectual property protections, your invention is not exclusively yours. This means anyone is free to independently copy, manufacture, and sell what you consider "your" invention. If there are no property rights assigned to a specific device, component, or process, there is no reason for a manufacturer to pay you for something they can design without penalty. Every invention must go through two simultaneous processes to become a product in the hands of consumers and generate income for the inventor. One is the intellectual property protection process (patents, trade secrets, and copyrights, for example), and the other is the commercialization process.

Risk Factors

The strategy of using potential costs to determine first steps after inventing something is called a **tiered-risk model**. **Tiered risk** is the amount of risk encountered at various stages of product development. To create a tiered-risk model, list the smallest risks of seeking commercial development and work your way to the greatest risks. Risks tend to become greater as development progresses, so evaluate each progressively greater risk, and determine which risks you can support and which you cannot. Without calculating your tiered risk, you run the risk of losing money from your commercialization and intellectual property protection efforts.

Seeking commercialization and intellectual property protection are complex processes that will vary widely depending on the nature of your invention and the industry to which it is targeted. When you have finished conceptualizing your invention, the first question to ask is whether to pursue a patent, develop a rudimentary physical model of your conception, or to approach possible buyers for your idea. Because your idea is largely unproven at this stage, the answer is: whichever strategy can be achieved quickest and costs the least amount of money.

A major consideration factoring into every inventor's decision is the ability to introduce his or her product into the market and the degree to which consumers will accept that product. Look for initial indications that your product is marketable by determining which industries might accept it, which products within these industries your invention might complement, and how well these existing products are selling. If initial market research indicates your idea will not be accepted or does not have an established market, you may not want to spend a lot of money creating a prototype or seeking a patent to own the intellectual property. If so, it may be best to put the invention on the shelf and wait until market trends open a wider opportunity for the idea. If initial market research indicates otherwise, think about patenting the idea and developing a prototype to show potential licensees.

The next step is to determine the cost of developing intellectual property protection versus the cost of manufacturing a prototype. At this stage, you have no guarantee a potential licensee will buy your idea, so if a product designer estimates a relatively low cost to make the prototype, it is better to spend money on creating a prototype before seeking a more expensive patent application. If the product requires a significant amount of money to prototype, potential licensees are more likely to want you to have an intellectual patent on the invention due to the money *they* would have to spend in order to produce it. If your initial market research indicates manufacturers are more likely to consider licensing if they see a prototype, you will have to consider spending money on making a prototype *and* applying for patent protection. A company is more likely to require a prototype if they need the invention's proof-of function demonstrated. For example, a company is far more likely to require a proof-of-function demonstration for a new type of theft prevention device for cars than it would for an uncomplicated invention such as a sticky note.

The Benefits of Licensing

Intellectual property rights are licensed or assigned rather than sold. In other words, when you transfer the rights of property to someone, you are the licensor. The individual or company that agrees to have the rights you own transferred to them is the licensee. Licensing is like renting or leasing your invention to a firm that will pay you for the right to develop it into a product for sale. If you permanently transfer intellectual property rights, you are not licensing the rights but rather assigning them. You cannot assign or license rights over to someone unless you were granted specific property rights to own the patent. You do not have to be an entrepreneur, manager, or business executive; you can keep being an inventor. Licensing a patent means you will be paid in exchange, partly in upfront fees and partly in royalties based on the volume of product sales. It also means you may not have to take on the risk of investing your own resources in product development, distribution, and marketing. The benefits of licensing include:

- The possibility of simultaneously licensing the same invention to a licensee in another field

- The ability to specify that the rights revert to you if the invention is not commercialized within a certain amount of time

- The opportunity to relicense the invention again in the future if the rights revert to you

- The ability to continue profiting from sales of your product without further effort on your part

- The opportunity to retain the right to the entire area of technology covered by your patent in case it covers valuable processes or products not yet invented

The risk of licensing is that you could license your product to a firm that may decide not to commercialize it. This most often occurs when there is a sudden change of leadership at the CEO level, and a decision is made to take the company in another direction that does not include your invention in their product line. You would still be paid, but would receive no royalties because of the company's decision to abandon product development. Based on the terms of the licensing you signed, you may no longer own the right to license or develop it elsewhere. Learning about contractual provisions can mitigate this risk. In a license, some or all of the intellectual property rights revert to the inventor under some conditions or at some time specified in the licensing contract. The inventor can then license the rights to others. If no rights ever revert to the inventor, then the contract is an assignment (or sale) rather than a license. Conditions under which an invention's rights may revert to the inventor under a typical license are if:

- The license expires after a certain time
- The licensee fails to commercialize the invention in a certain time
- The licensee abandons the invention and ceases to manufacture, distribute, or sell it
- The licensee goes bankrupt
- The contract or provisions of the contract are violated
- The licensee fails to pay royalties to the inventor

So, what makes the rights to an invention attractive to a potential licensee? Typically, someone sees the commercial potential in licensing your intellectual property when there is a definable market for your product and when that product that can be created, distributed, and sold at a profit along a distribution chain. Licensees also like some sort of exclusivity granted through your intellectual property rights because it means no one else has the right to design and ultimately compete with the idea they intend to convert into a commercial product. If you can convince a manufacturer your idea is both marketable and unique, in most cases, a manufacturer firm will be interested in purchasing those rights.

Your Presentation

By granting licensees the right to make and distribute your invention, not only are you free from covering the associated costs, but you also are free to continue inventing other products. If you cannot find a licensee or buyer, or you want to retain ownership, your remaining option is to build a business around your product, pay a manufacturer to build it, and recruit a distributor to sell the product to stores. In either case, both of these options require the inventor to enter into a set of partnerships. To acquire them, solicitation is necessary — more so for independent inventors with limited access to the resources required to bring a product to market. To solicit a licensing agreement, the materials you need for your presentation should include:

- Background information about your prospective licensee: history, product lines, market position, executives' personalities, and the like

- A written description of your invention: what it is, what it does, how it works, whom it is for, and the benefits the consumer will realize from using it

- A professional-looking, one-page summary of your invention, the results of your market research, your intellectual property (IP) position, your recommended price, and revenue forecasts

- Documents related to IP protections you have obtained for your invention

- Suggestions regarding distribution and retail channels that complement how the company already works

- Artwork detailing how the final version will look

- A working model or prototype

- Sample packaging for your prototype

- Information from other manufacturers or consultants about manufacturing costs and product specifications that provide enough detail for the licensee to estimate likely manufacturing costs

- Illustrations or a mock-up of packaging ideas

- A marketing pitch that addresses:

  - Your product and how it is better than existing products

  - Responses to anticipated objections

  - Information about distribution channels

  - Highlights of how your invention complements the licensee's market position and existing product lines

  - Information about the size, affluence, and demographic trends of the consumer market for your invention

- Financial figures estimating the retail cost, the manufacturing cost, the profit, and the length of time it will take the licensee to recoup its investment

- A draft license or suggested terms

Begin the meeting by establishing rapport. Talk about your background, citing whatever related credentials you may have in the industry, as well as what influenced you to create this invention. Have a pitch memorized or outlined, but do not read a proposal off cue cards. Establish and maintain eye contact, smile, and let your passion show. Highlight the benefits the end user will value, how the invention complements the firm's positioning and product line, and demonstrate where your market analysis indicates how profits can be attained. At the end of the pitch, showcase the prototype, and hand it to the audience.

Be prepared to answer questions regarding your research as it relates to expected manufacturing costs, retail price, profit potential, sales projections, and competition. If you have applied for a patent, you may wish to use language from the specifications section. You may also wish to bring a copy of the specifications from the application form in case the customers insist on seeing it. Only provide copies of your presentation after you have finished, and even though a nondisclosure agreement should have been signed at this point, do not leave any written material, drawings, or diagrams with your prospective customers. The entire pitch should take no more than ten minutes. Once you have finished, thank them for their time and allow for questions. The questions asked and responses given will reveal their initial thoughts about your invention, their priorities, and what conditions may affect decision making.

At this time, be open to any suggestions your potential customers may make. Remember, it will be their product, so encourage them to take this kind of ownership. If the prototype has been designed for functionality, you will likely be asked for a demonstration of its use. If use of the prototype cannot be applied in that setting, prepare a video demonstrating the prototype in the setting where your product normally would be used. When the Q&A session is over, prospective customers will either make their decision on the spot or convene to discuss it and inform you of their decision later. If they express interest, begin to explore what sort of deal they might be interested in making. If they reject the presentation, ask for feedback regarding what they believe to be your invention's strengths and weaknesses. The meeting should end with you soliciting referrals to other prospective licensees. In some cases, potential customers may reject your presentation because they are moving away from the existing market space they command. If they do not see your product as direct competition to any of their existing or future product lines, they are likely to provide information regarding their competitors.

If the prospective customers remain interested, discuss the scope of your intellectual property protection and what rights you are willing to license or assign. If they start talking about licensing rather than assignment, follow this by asking what minimum number of sales they would be willing to guarantee in a deal. If the prospective customers acknowledge their interest in striking a deal, present them with a drafted license agreement stating your preferred terms. Do not expect them to sign anything immediately, as most companies will send agreements through their legal department first to make sure the language and terms have been examined. If so, provide a date by which you will need to hear from them.

The prospective licensees may say they need to evaluate your idea or perhaps even test your prototype. Any prototype you provide will be handed to the engineering department. Marketing research provided will be assessed by the marketing department (though they will conduct their own study). Most companies that routinely sign licensing deals will have an estimate as to how long a response should take. Before pitching them, contact their previous licensors, ask how long the company took to provide a response, and use that as a benchmark for how long you are willing to give them. If they suggest a week to ten days, that may be reasonable. If they want to hold your product off the market for longer than that, they should pay for the privilege with an option fee. An **option fee** grants the right to "rent" a product for a certain period time in return for your willingness to potentially lose business opportunities. If you accept an option fee, you should not solicit other companies during the time the option fee is in effect. The amount paid in the option fee should always be commensurate to the length of the term. If the company is seeking an option, they will deduct whatever has been paid in the option from the first royalties should they choose to license the invention. If they do make an offer and present their own licensing agreement, you may not wish to sign anything immediately either. Accept the document, and say that you will have an attorney review the terms. If they are acceptable, you will be in touch.

Ongoing communications with a prospective buyer

Regardless of how long your invention is kept in the hands of a potential buyer, be sure to a keep a record of the dates, conversations, and agreements that have been exchanged or agreed upon. Documentation of all transactions ensures any drawings or confidential documents exchanged will be returned and provides evidence that you solicited business to them. Establish a paper trail as your communications advance. After each phone call, send and save an e-mail thanking them for meeting with a brief recap of your understanding of the meeting and the next steps. These files can help you in court if you feel your invention has been copied.

Licensing Agents

A **licensing agent** or representative provides contracted marketing services to get your invention in front of appropriate prospective licensees in return for a percentage (anywhere from 10 to 50 percent) of your total gross licensing deal. The advantages of having a licensing agent include:

- Professional representation. Large corporations and businesses in some industries prefer to work with inventors through agents who understand the executive process of licensing.

- Confidence. Some inventors do not feel they can give a convincing pitch or presentation.

- Time. Some inventors are too busy and prefer to pay someone else for the legwork.

- Access. Agents have large networks of professional contacts in trade associations and can gain access to decision makers better than independent inventors without trade membership or industry contacts.

Licensing Elements

Licensing agreements have several elements, and there are several kinds of agreements. The most straightforward kind of agreement is an exclusive licensing agreement. In an **exclusive licensing agreement**, the licensee acquires the sole right to exploit your patent claims, and no one else can acquire them. If the invention can capture market share in more than one geographic area, distribution channel, or industry, it may be best to obtain a nonexclusive license. In a **nonexclusive license agreement**, you retain the right to license your patent claims to more than one licensee. For example, multiple manufacturers and distributors may dominate one region but have little market share in another. In this case, you want to offer a second license agreement to the manufacturer of that second particular region. The elements of any type of licensing agreement may include:

Payments

- The amount of the payment up front
- Whether the upfront payment is deducted from future royalties
- The royalty rate
- When the inventor will be paid the upfront payment and at what intervals the royalties will be paid
- How the inventor can audit or check the veracity of the sales reports provided by the licensee
- The way royalties are calculated, i.e. gross sales or net sales. (Never accept any other basis for calculations, such as profits, or allow deductions for expenses. These soft numbers can be fixed against you.)
- What recourse the inventor has in the case of overdue payments
- Whether the licensee wishes to retain you to provide consulting services, at what rate, and for what minimum number of hours, if any

- Deductions. The customer may try to shift the manufacturer's costs by getting you to pay for some items out of your royalties. You are better off having a slightly lower royalty rate than the uncertainty of the size of these deductions.

Rights

- The scope of the rights being conferred — a patent, trade secrets, or copyright

- The minimum sales performance below which the rights will revert to the inventor

- The time period of the license (If there are few or no intellectual property protections on the invention, time limits do not mean much because the invention will be in the public domain. If there are intellectual property protections, you can capitalize on them by requiring a renewal of the contract a certain number of years out.)

- The geographic scope of the license (United States rights only, or international)

Legal protections and responsibilities

- Whether the licensee will pursue patent protection in the inventor's name

- Whether the inventor or the licensee will be responsible for pursuing infringers (This is an enormous legal liability; make sure you are never saddled with it, as it could bankrupt you.)

- Product liability (The manufacturer is held responsible for product liability by those who purchase your invention products. Do not let the licensee pass this liability on to you. The licensee may have deep pockets; you may not.)

Assigning Rights

Assigning rights to an invention means permanently selling your intellectual property rights. Assignment is a more attractive option than licensing when you need more cash than an initial upfront payment from a licensing agreement can provide. Assignment rights also shift the litigation risks to a firm that is going to commercialize the invention. In most cases, inventors choose to license rather than sell their rights for a variety of reasons. By licensing, the inventor will continue to profit if the product continues to sell. Also, the rights can revert to the inventor if the licensee does not commercialize the product (if the license contains these conditions). In addition, the inventor may be able to license the invention to firms in other fields simultaneously. The area of patent protection may cover new, profitable technologies not yet invented. It may be easier to convince a company to pay the minimal upfront cost of licensing and reduce its risk of spending money on an invention that flops.

Once you make money on your invention, the IRS requires a Schedule C or Schedule E be filed with your tax return to deduct all expenditures made to create your invention. The IRS will only accept these deductions if you can provide a full accounting of expenditures through documentation. It is important to save receipts on prototypes, evaluation services, and other materials.

The Negotiating Process

The first step is to rank your negotiating terms in order of importance and maintain at least two or three non-negotiable items. Reject any requests to negotiate those items, but counter with something you are willing to give up in return. Counter it with an offer of your own that gives your customer what they want and claims something you want in return. Remember, a licensing deal with an upfront payment and royalties that shift the risk of commercialization to your customer is an excellent deal. The most important non-negotiable terms may include seeking upfront payment, a competitive royalty rate, a minimum guarantee, and details as to whether the royalty rate will change after a certain number of units are sold.

Begin negotiations from a strong and confident position, but engage with your prospective licensee in a flexible and easygoing way. If you have one or two suggested changes in the terms, rewrite the document, sign, and return your version. Doing so puts the onus on the potential customer to identify and decide whether to accept the terms you have negotiated.

Some elements in every agreement should not be subject to negotiation. For example, a manufacturer should never ask the licensor to take on their liability or be made responsible for pursuing infringers. Recognize what terms you are willing to concede and what terms will not render a favorable agreement. During the negotiating process, it is important to maintain the affable demeanor. View your prospective customer as a partner because having a partnership means you now have common interests.

So, what terms should be included in your contract? First-year earnings can be a basis for calculating an upfront payment (perhaps 20 to 30 percent of annual earnings). If you do not learn about the standard royalty rates through a patent attorney or key decision maker during market research, consult with other inventors in your field who can provide information about typical royalty rates. You may learn things during the process of negotiation that may cause you to elevate or lower your target number before you disclose it to the customer. When asked, you should be able to respond with figures you desire, but can afford to discount during negotiations.

Profit equals retail price divided by two for each juncture in the distribution chain between manufacturer and customer minus manufacturing and materials cost. For example, let us suppose that you have invented a specialty bag for some type of photographic equipment. You have a prototype and the prototype manufacturer has told you that a large run would bring the price-per-item down to $3. You have already examined other bags and estimate a retail price of $27 for your bag. Within the distribution tier, there is one manufacturer, one distributor, one wholesaler, one retailer, and the end-users. The retailer can charge $27. That means the wholesaler has charged them $13.50, which means the distributor has charged them $6.75. The manufacturer can charge $3.38.

You would like to make between 5 and 10 percent of the price your licensee charges. At 5 percent, your royalty per item is just under 17 cents. This leaves the manufacturer a profit of 21 cents on each item, a tad over 6 percent. If the manufacturer believes it can charge more retail or make the item cheaper through its resources to increase profit, it might be interested in licensing your product.

Once you know how much your customer stands to profit by your invention, you can estimate how much they might be willing to pay to license the invention. Your market research will enhance your ability to project your invention's life cycle, as well as the number of units sold per year. The royalty rate paid inventors on average ranges between 5 to 15 percent of net sales.

According to Steve Overholt of the Gannon University Small Business Development Center, "The royalty percentage that [a licensee] will pay is related to the total value contribution that your invention makes to a product. If your invention essentially defines the entire product, then you can command a relatively high royalty percentage." Different industries have differing royalty rates licensees are traditionally willing to pay. One source for identifying royalty rate is RoyaltyStat® at: **www.royaltystat.com**. This data shows that the vast majority of royalty rates are between 2 and 10 percent. For a subscription fee, this site and others can provide more detailed reports by industry.

Inventors should insist royalty payments be calculated based on **net sales** (all sales minus returns or nonpayments). This is the most easily established and verified number and is a basis that makes sense in terms of the inventor earning a portion of what the licensee earns. Licensees who calculate inventors' royalties based on profits can inflate costs to decrease apparent profits, and it would be difficult for the inventor to find out or prove.

Based on your figures, how long will it take the manufacturer to cover its own startup investment costs at various lump sum plus royalty figures? If it can make its money back relatively quickly, you may pique its interest. What sort of upfront payment should you request? Determine what you think is a realistic figure the company might be willing to pay. Make an offer somewhat higher than that figure, and negotiate from there.

Ask for close to the standard industry royalty rate to demonstrate your knowledge of the industry and create a professional appearance. Expect to end up with the standard rate. You want to maximize your income but not drive off your prospective customers. They need to perceive you as reasonable, realistic, and well-informed about their industry. If you come off as unreasonable, greedy, or uninformed, they will not feel comfortable doing business with you. You might ask for one royalty rate on the first 10,000 items and a slightly higher rate on all items afterward. The customers may be more willing to negotiate the later rate because they may have covered their investment by that time.

Contingent Fee Intermediaries

Alternative strategies exist for getting a product to market. Some inventors use **contingent fee intermediaries** — experts trained to find suitable manufacturers and distributors for an invention. These brokers can also act as an agent in the transfer of your property rights. They can help market the invention but should not be confused with industrial developers. Rather than helping you develop a market strategy and prototype, contingent fee brokers will work to market and license the invention for a percentage of the intellectual property rights. Most are considered reputable; however a small percentage of brokers qualify as scam artists. *See Chapter 10 for* *more information.* Many of the trusted contingent fee brokers can be found through inventor associations. Some include:

- **America Invents (www.AmericaInvents.com)** helps license to companies, with products sold in more than 50,000 outlets nationwide and more than $1.5 billion in retail sales.

- **Big Idea Group: (www.BigIdeaGroup.net)** an open-source network of 13,000 problem solvers and inventors available to solve inventor's innovation challenges

- **Inventors Publishing: (www.InventorsPublishing.com)** intellectual property litigation lawyers offering agent services, consultation on e-commerce, data compression

When dealing with contingent fee intermediary brokers, you should not be asked to pay an upfront fee, as this is typically the sign of a scam. To avoid accounting issues, all negotiated royalties should be sent directly to your address. Some inventors hire brokers to find them licensing deals and then handle the negotiating themselves. If your product requires a relatively uncomplicated manufacture, it may not be necessary to find a broker or a licensee. Instead, contract a manufacturer located in China, and choose a U.S. distributor. Although you may have the additional headache of having to pay manufacturing costs, supervise the process, and assume all the risk, the reward for success is 100 percent of all profits.

Assistance for Independent Business

If you find that prospective licensees do not feel confident enough about the size of your market, you may have to demonstrate otherwise by creating a small business to penetrate one or two test markets. To accomplish this, an inventor may seek investors or partners willing to help bring the product into a test market. Investors or business partners can increase both the amount of money contributed to the startup and the business savvy of a new startup. They should be individuals who bring contacts, capital, experience, and expertise to areas that are critical to the growth of the startup. They should be people the inventor is willing to work closely with.

As you seek funding, be aware of the difference between a loan and an investment. A loan must be paid back according to the terms on which it was lent. An investment provides the investor with partial ownership of your enterprise. The U.S. Small Business Administration (SBA) is a great resource for entrepreneurs to begin looking for funding. The SBA has the influence with banks and offers loan programs for small business owners. To find out more, visit **www.sba.gov/category/navigation-structure/loans-grants**. Banks may be willing to fund your startup without a guarantee from the SBA, though their allocations have grown tighter in recent years. One way to attract a financial institution with lending capital is to market your product without producing a commercial run and use the orders as leverage to obtain financing from a bank or obtain a line of credit from a manufacturer. This means marketing your product before it is produced to generate orders to be fulfilled later. Running your own business means the distribution and marketing is now your responsibility. Use market research to find the appropriate industry-related distribution channels. Your customers are no longer the manufacturers, but rather, the middlemen — warehouses, distributors, import and export firms, and others.

Another type of investment available is through **angel investors** — affluent individuals who capitalize startups in exchange for convertible debt or ownership equity. Angel investors typically expect a 20- to 50-percent annual rate of return on their investments and insist they be indirectly involved with the business. Be aware that businesses soliciting investments must follow U.S. securities laws. For more information these laws, see **www.sec.gov**.

When angel investors and financial institutions are not interested, inventors turn to their family and friends to borrow or solicit investments. Funding from family or friends should only be used for product development because this is the most speculative phase of your business, the part most unlikely to qualify for loans from other sources. Another source of potential funding is a venture capitalist. **Venture capitalists** are institutionalized versions of angel investors. They are, however, far more selective in their investment choices and expect an annual return rate of about 40 percent on their investments. Another source of investment is through federal funding. Eleven federal agencies have Small Business Innovation Research programs that fund inventors to commercialize research in the agencies' areas. They include:

- The Environmental Protection Agency
- The National Institute of Standards and Technology
- The National Institute of Health
- The National Science Foundation
- The Departments of Education, Defense, Transportation, and Energy

To find out more about the SBIR program, go to **www.sbir.gov**. The SBIR program has state affiliates that can help inventors locally. A list is available at **www.sbir.gov/state**.

When you build a business, you may sell your product directly, or you may work with marketing representatives who work on commission and present your product to the retailers they have developed relationships with within their industry. If you have not found a mail-order catalog to carry the invention, consider developing one with an online marketing firm. They can provide an avenue for inventors and entrepreneurs to link with their prospective consumer markets.

If you have created an invention with someone else, you have joint ownership. Joint ownership has four forms: ownership as a result of co-inventing, ownership created by investment, ownership created by legal transference, and ownership created by contribution of expertise. In a joint ownership, all owners must consent to the assignment of rights.

If you acquire a joint owner through investment, make sure the investor shares the same vision and business interests for the invention that you do; otherwise, disagreements over licensing deals can hamper the entire process. An agreement signed up front designating the rights of each co-inventor will prevent misunderstandings and hard feelings down the road.

Before opening a business and seeking investors, consider whether you really have intellectual property rights to your invention. Inventors who work for universities and companies are often bound by an agreement whereby the employer or university owns the right to your inventions. If the inventor was hired in a "work for hire" arrangement, specifically to develop the invention, then the rights to the invention belong to the employer. Even without such an agreement, a company may have the right to use your invention without any additional compensation if you worked on the invention during work hours, at the work site, or used employer materials. A major source of revenue for universities comes from royalties earned by employees in research and development programs. Under the Bayh-Dole Act, universities, small business, and nonprofit organizations control the intellectual property of employees operating with internal or federal funding. Royalties go through the university, and a percentage is paid to the inventor. If your employer owns the rights to your invention and is willing to transfer the rights, they will be required to sign a letter confirming ownership of intellectual property rights. Included in this letter should be a confirmation statement, description of the invention, date, and signatures of employee and employer.

Perhaps the most famous example of a university's right to exercise control was University of Florida versus employed researchers who were asked by the university to develop a formula that would quench the thirst of their athletes. The formula they came up with was Gatorade®, named after the team itself. When the inventors took the product beyond to the greater market, the university demanded and received 20 percent of the revenue that the inventors earned after securing outside licensing deals. To this day,

Gatorade is the official drink of many university sports programs, and the University of Florida receives about $8 million annually from licensing.

Types of manufacturers

If you plan to supervise the manufacturing process rather than license to a manufacturer, you will have to know what manufacturing methods will produce your product. Manufacturing categories include:

- Molding
- Mold making
- Casting
- Forging
- Chemical
- Machining
- Electronics
- Textiles
- Die making
- Wood working
- Wire forming
- Extrusion
- Stamping
- Engraving
- Fabrication
- Assembly
- Packaging

Original equipment manufacturers (OEMs) are the firms that create the parts assembled into or used to produce a finished product. To supervise manufacturing, you may need parts from several OEMs and the services of an assembly firm. Search for American manufacturers by state and type in the *Thomas Register*, available at library reference desks, or online at

www.thomasnet.com or at **www.mfg.com**, an online resource community for locating parts specialists and manufacturing companies.

If you have worked with a designer to create a model and determine what type of manufacturing process best suits your invention, you may have received references for suitable manufacturers. Your designer will have recommended the type of tooling necessary to manufacture your invention. Tooling is a manufacturing term that refers to manufacturing aids used in the limited or specific production of something. Types of tooling include jigs, molds, patterns, dies, cutting tools, gauges, etc. Each part of your invention must be created using some tools. Some parts can be made using standard or generic tools, while other parts require the creation custom molds or other custom tools.

Manufacturing large runs of your product requires more precise, refined, sturdy, reliable, and expensive tools. Different manufacturing processes will have different tool requirements, which can have significant implications for the cost of manufacturing your product. If you are planning to work with a manufacturer who will be creating and using custom tools, find out how they handle the transfer of tools if you elect to switch firms down the road.

If you are contracting for an ongoing relationship, some manufacturers may be willing to prorate the cost of the tooling into the cost-per-unit of your products to help you distribute the cost of the tooling over time rather than having to come up with all the capital at once. Make sure you feel comfortable working with a manufacturer and their personnel because you will have to supervise their work and accept the results. Asking manufacturers for references can help you avoid expensive learning experiences.

When contacting their clients, ask how satisfied they are, what issues have arisen, and how the manufacturer handled them. Try to ascertain which clients have similar products made by the manufacturer so that you are not paying for a learning curve. If you are working with several manufacturers and an assembler, you will need to look at the scope of work for each contract carefully to ensure that the process will go smoothly. You do not want to be stuck in a situation where the Original Equipment Manufacturer (OEM) had sent a product that is not finished enough for the assembler to use.

If the manufacturer delivers poor quality goods to you or your assembler, you must have some recourse, so make sure the contract is specific about quality guidelines, tolerances, margins of error, and methods of recourse. Expect to work closely with your manufacturer while the manufacturing process is getting underway. Try to maintain good working relationships with your manufacturer; you have a lot invested in making the relationship work. If you are working with a foreign manufacturer, monitor currency exchange rates, the price for international delivery (including taxes and tariffs), and translators. In addition, you will have to comply with U.S. Customs and Border Protection regulations, which can be found at **www.cbp.gov**.

New businesses must struggle to balance the price breaks afforded by manufacturers for larger runs or orders with the cost of storing excess inventory. They can have cash-flow problems, and new products usually require marketing. Know the amount of time your manufacturer is going to need to ship new products, and market accordingly. New businesses are much more likely to fail because of poor cash-flow management than from a lack of inventory. Tying cash up in excess inventory is not the best use of capital for a new business.

CASE STUDY: STRIKING A LICENSING DEAL

Kim Babjak, president
KimCo, LLC
P.O Box 7641 Loveland CO 80537
kim@kimbabjak.com
www.kimbabjak.com
Phone: (970) 663-2382
Fax: (602) 926-0443

I was working the day shift at a local fast-food restaurant when I brainstormed the idea for my first invention, Zip-A-Ruffle®, which I named to describe the product as being an easily interchangeable bed skirt. From the beginning, it was an idea I thought had a lot of potential on the bedroom supplies market. I wanted to retain control of the manufacturing and distribution process, so I learned how to source materials, find manufacturers, import finished goods, and submit inventions to distribution channels and retailers. It took me three years to get my first invention to market.

The market research I did showed that QVC's customers were good matches for the ruffle, and I knew how they worked internally, so I submitted it to them. After I was able to convince QVC to carry it, I was able to approach the larger stores such as Walmart, QVC UK, Walgreens, and the bigger box retailers. Eight years and several million dollars later, I have commercialized a line of lingerie and built a robust business on a number of different inventions carried by retail stores. I have taken what I have learned and expanded from commercializing my own inventions to supporting other inventors in commercializing their products.

Not only is market research key to getting your product out there, but also it is essential to knowing what type of patent application you should seek. After I was denied a patent for the Zip-A-Ruffle, I eventually realized that it is better to file a provisional patent, which protects your idea while giving you time to figure out if the idea is going to work or not. You have to be willing to let go of ideas that are not going to make you a profit and move on to those that will.

The money you save on not having to file applications for full patents is better put toward getting a patent attorney who can bring the product right to the decision makers without delay. Independently submitting to QVC is difficult and does not lead to responses very often. To find the right licensee, you have to locate several, interview them, and make an informed decision. Commissions vary from 10 to 15 percent. The good ones get 15 percent, so when it comes to representation, do not let that stand in your way.

Patent Protection

Liability

A patent owner may sue any manufacturer who makes, offers for sale, sells, uses, or imports a patented device or practices a patented process covered by claims. A patent owner also may file suit against a retailer or the purchaser of a patented invention. The theory of vicarious liability provides that a business be liable for infringement by its employees or agents under the following conditions:

- The agent acts under the direction or authority of the business entity.
- The employee acts within the scope of his or her employment.
- The business benefits from, approves of, or adopts the infringing activity.

Losing patent rights

Patents provide protection for an invention for as long as certain requirements are met. If those requirements are not met, patent rights may be lost. The USPTO may issue a patent for an invention. However, the patent may not provide broad enough coverage to adequately protect the invention being challenged. The following situations can cause you to lose patent rights:

- The patent owner does not pay the required maintenance fees for the patent.

- It is determined that the patent fails to describe the invention adequately.

- It is determined that the patent fails to define adequately how to make or use the invention.

- It is determined that the patent includes inadequate claims.

- The discovery of prior art references indicates that the patent is not novel, or it is obvious.

- The patent owner engages in illegal conduct connected with the patent.

- The patent owner does not disclose all information or material to the PTO.

Infringement

Infringement in the United States occurs when a party other than the party who holds patent rights to an invention intentionally sells, makes, offers for sale, or uses a patented invention in the United States, any U.S. territory or any U.S. possession, or the party imports the invention into the United States during the term of the patent. Infringement also occurs if all the parts of a patented invention are created and shipped to a foreign nation with instructions for the assembly of such parts.

Infringement can only occur during the life of a patent. As such, a patent actually must be issued for infringement to occur. Infringement cannot be argued during the pendency period of a patent application unless the patent application is published before the patent is issued. The patent owner may sue for royalties from the date of publication, provided a patent is later issued, and the infringer has actual notice of the published application. The patent owner must mark patented devices with the patent number assigned by the PTO. If a patent owner fails to provide such a marking, damages only may be recovered from the date on which the infringing party was notified of the infringement.

A patent owner may authorize the use of a patented invention. However, any attempt to extend that authorized use beyond that for which it was intended is a form of infringement. A patent owner, for example, may authorize a company to build and use a single patented device. If instead of making a single device, the company decides to build and use three such devices, the building of the additional devices constitutes an infringement.

It does not matter whether the infringing party intentionally or inadvertently develops a patented invention; the unauthorized sale, use, or manufacture of the invention is infringement. The intent and knowledge of the infringer have no bearing on the act of infringement but may be used in determining the amount of monetary damages and the level of contribution to an infringing act.

Infringing patent claims

To qualify as infringement, a device or process must physically perform or possess all of the elements documented in at least one of the claims specified for a patented invention, or the infringing device or process must meet requirements specified in The Doctrine of Equivalents. Each claim of a patented invention should define a different and distinct area of protected technology. If a device or process infringes on one or all of the elements in a single claim, that is all it takes to be considered infringement.

If a claim includes four elements, a process or device must contain all four elements of the claim to infringe on the patented invention. If the process or device includes three or fewer elements of the claim, no infringement occurs. If an invention is supported by a dependent claim, a device or process must infringe upon the associated independent claim of infringement to occur. Dependent claims will not be considered infringed upon if the independent claim is not infringed. Also, the violating party does not have to make use of the totality of the patented invention. If the violator infringes on the essential parts of the patented invention, the violator may be guilty of infringement. When a device or process infringes on at least one claim in its entirety, it is considered a direct infringement. When someone persuades another party to infringe or contributes to the infringement of another, it is considered indirect infringement.

Direct infringement

A party commits a form of direct infringement, termed **literal infringement**, when he or she directly makes, uses, or sells a process or device that contains every element of a patent claim. With literal infringement, the infringing process or device is a mere imitator of the patented process or device. In many instances, the infringing party attempts to capitalize on the success of a patented invention by capturing a market share of the proceeds from sales. A party commits a form of direct infringement, termed **equivalent infringement** or **infringement under the doctrine of equivalents**, when he or she designs a process or device around a patent claim such that it functions in substantially the same manner and provides the same result as that of the patent claim.

Infringement under the Doctrine of Equivalents

The Doctrine of Equivalents is a judge-made rule that establishes infringement based upon the similarity of function, method, or results. The use of the doctrine has been met with controversy, and its application has been curtailed by virtue of a decision handed down by the Court of Appeals for the Federal Circuit. The Doctrine of Equivalents requires that every element of a claim be literally or equivalently present in the infringing device or process. **Equivalents** are elements of a device or steps of a process that were either available when the patent was issued or available after the patent was issued but before the patent was infringed upon. The doctrine prevents an infringer from designing around patent claims by making minor alterations or using later developments that were not available when the patent application was filed. An element of little or no importance to the claims of an invention can have a wide range of equivalent elements that are not disclosed in the specification of the patent application.

Means plus function

When a patent claim specially describes a function of an element rather than the structure of the element, the claim consists of a means-plus-function clause. In other words, a **means-plus-function** exists when a claim specifies a particular way to perform a function without describing how to perform the function or with what to perform the function. When the scope of a claim includes language that considers "means-plus-function," the means of performing the function is considered as described in the specification and its equivalents. The clause usually begins with the term "means." A hinge, for example, may be disclosed in a patent claim as, "a means of attaching a door to a cabinet," without indicating the type of device necessary to do so. In a literal infringement case, the court would have to determine equivalency or whether the means described in the specification for the patented invention was the same or equivalent to the means of the infringing device. Even when the infringing device is not a literal copy of the patented invention, infringement may still occur if the device or process performs substantially the same function, in substantially the same manner, and obtains the same result as the patented invention.

The existence of the term "means" in a claim will not always be grounds for an equivalence analysis. If the means of a clause is specific and the infringing device meets the means clause, the mean-plus-function equivalency is unnecessary. If the means of an invention discloses a physical structure that is insignificant to the claimed invention, many more equivalent structures may exist than would be the case if the characteristics of the structure were critical in performing the claimed function.

As an example, if screws were the preferred embodiment and means of holding a cabinet together, then nails, dowels, or other types of joints would constitute equivalent elements because the screws are not critical to the functioning of the cabinet.

An equivalent structure or act cannot embrace technology developed after a patent is issued. The comparison between what is disclosed in the patent specification and what is proposed as an equivalent must consider the overall structure or process. The infringing device or process may have fewer or more parts or fewer or more methods, but it functions in substantially the same way as the patented device or process.

If an element of a patent claim is amended during the patent application process, the claim is presumed narrowed to bar the doctrine of equivalents. The presumption can be challenged if the patent owner is able to demonstrate that the amendment involved a feature of the invention at the time of application or for some other reason could not be included in the original claim. In other words, the doctrine of equivalents may still apply in an infringement lawsuit when a claim has been amended, provided the patent owner is able to overcome the presumption of the claim.

Indirect infringement

Indirect infringement occurs by contributing to an infringement or persuading another party to engage in infringement. An indirect infringement cannot occur unless a direct infringement occurs. There are two ways to infringe indirectly upon a patent.

1. **Inducing Infringement**: A party is persuaded to make, use, or sell a patented invention without authorization.

2. **Contributory Infringement**: A material component of an invention is sold with the knowledge that use of the component is unauthorized.

A party who sells infringing parts of an invention has not engaged in the act of infringement until or unless those parts are used in an infringing invention. A typical example includes a manufacturer that knowingly sells a patented device that serves no other use except as it is patented. When a buyer makes use of such a device, the manufacturer is considered a contributory infringer.

Infringing a design patent

Infringement upon a design patent differs from infringement upon other types of patents. The scope of rights provided by a design patent depends upon the drawings that accompany the patent, not the claims of the patent. Infringement is assessed by whether the appearance of the infringing design is substantially the same as the design claimed by the drawings of the patented invention.

Design patent infringement is measured by the Gorham test. The Gorham test determines whether an ordinary consumer finds the patented and the infringing design to be similar such that their resemblance is considered deceptive to the consumer and induces the consumer to purchase the object containing the infringing design. In order to administer the Gorham test, the party suing for damages must be able to demonstrate evidence that ordinary observers and purchasers are being be misled by the apparent "sameness" of the two products. In the ruling of *Gorham vs. White*, the Supreme Court ruled that "if, in the eye of an ordinary observer, giving such attention as a purchaser usually gives, two designs are substantially the same; if the resemblance is such as to deceive such an observer, inducing him to purchase one supposing it to be the other, the first one patented is infringed by the other." The two designs must not only be similar, but also the infringing design must contain the novel features of the patented design. Novel features include those design features of a patented article that distinguish it from prior art.

The patent may not be used to protect the functional features of the article for which the patent is obtained. A design patent, for example, may protect the unique design attached to an article of clothing. However, a design patent does not protect the manner in which the design is attached. The most common defense in a case of design patent infringement is that the patented design lacks the requirement to be ornamental and, thus, is an invalid patent. The argument is usually based on the contention that the design is of no concern to consumers.

Infringement Hearings

Infringement, for other than design patents, is determined as it relates to the scope of coverage for the infringed-upon claims. The language used to document a claim in a patent application defines the scope of the claim. As such, the language of a claim is open to interpretation of the reader. Both the patent owner and the infringing party are allowed to present their interpretation of claims to a judge in what is termed a Markman hearing. This hearing is held before taking the issue to trial. During the Markman hearing, the judge decides the scope of each presented claim. If the language of a claim is ambiguous, the judge may use the description, drawings, and if necessary, the entire patent application file to interpret the claim. The judge may also receive expert testimony to assist in interpreting the claim.

Court hearing

Infringement is a matter that must be brought before a federal court except when the infringing invention is being imported into the United States. If the invention is being imported, the patent owner may also bring proceedings before the International Trade Commission to halt the transport of the invention at its port of entry into the United States. The federal court hearing is located at the federal court in the district of the defendant's residence or the district where the defendant committed the infringing act and has a regular and established place of business. The defendant is the infringing party.

The patent owner must decide which federal district court is in the geographic location, known as venue, appropriate to handle the particular case of infringement. It is common for patent owners to sue a retailer or customer to establish venue closest to his or her own residence. Manufacturers have the burden of defending customers and retailers in infringement lawsuits regardless of whether the manufacturer is local to the residence of the party they are defending. If the defendant is a corporation, the residence of the corporation is any district where the corporation is incorporated, doing business, or is licensed to do business. If the defendant is the federal government or a contractor under a federal contract, the lawsuit must be brought before the U.S. Court of Federal Claims in Washington, D.C. Patent owners are entitled to sue the government for infringement and receive compensation for loss of use incurred as a result of use by or for the government.

In 1999, the Supreme Court ruled that under Constitutional principles, states could not be held liable for patent infringement. The plaintiff in an infringement lawsuit must prepare three particular documents and file them with the appropriate federal court. The federal court's rules of service and process dictate that a copy of each document also must be delivered to the defendant.

1. **A complaint**. The complaint sets forth the facts of the infringement and requests remedies, such as compensation or an injunction. If the plaintiff desires a jury trial, that request must be included.

2. **A summons**

3. A civil cover sheet

The defendant must file a response to the complaint in which the defendant denies or admits the accusations and provides a listing of defenses to such. If the plaintiff did not request a jury trial, but the defendant would like to have such a trial, it must be requested with the answer to the complaint. If the defendant wishes to file a counterclaim against the plaintiff, the counterclaim also must be filed at the same time that the answer is filed. If an act that is cause for a counterclaim arises out of the same occurrence or transaction that is the subject of the complaint, it must be filed as the counterclaim. Such a counterclaim is compulsory. As an example, a patent owner commits battery against the company CEO that he believes is infringing on the patented invention. If the patent owner then files an infringement lawsuit against the company, the company CEO must file a claim for damages as a counterclaim in his or her answer to the patent owner's complaint. An appropriate federal district court will examine the claims of the patented invention against the infringing device or process.

A patent owner may request that the court issue an injunction to prevent any further infringement of the patent and request an award of monetary damages for the infringement. Should a party be found guilty of infringing upon the rights granted by a patent, the patent owner is awarded damages to compensate for the loss incurred by the infringement. Damages for design and plant patents generally are awarded at reasonably assessed royalty rates that are based on the amount of loss in profits due to the infringement. The court, for example, may award damages as the amount of royalty payments the patent owner would have received had the patent owner licensed the sale of the infringing device. If the infringement is found to be deliberate and in bad faith, the amount of reasonably assessed royalties may be tripled. This is known as **enhanced damages**.

The infringer may be ordered to pay the patent holder's legal fees. When the infringing party is the federal government or a contractor to the federal government, compensation is limited to damages plus interest.

The infringing party may be issued an injunction that specifically orders a halt to making, selling, or using the invention. A court order may halt the manufacture and sale of a patented invention even if the invention has not been properly marked with the patent number. The patent owner may request a preliminary injunction that halts the immediate activity before a trial begins, a permanent injunction that is granted at the end of a court trial, or a temporary restraining order that halts the activity for a short period.

When a temporary restraining order or preliminary injunction is ordered, the plaintiff is required to post a bond, which is intended to cover costs and damages should the defendant prevail in the lawsuit. Temporary restraining orders may be issued without notice to the infringing party if it is concluded that immediate damages will result from the infringing activity. As an example, the court is convinced that evidence of an infringement would be destroyed. A temporary restraining order is issued and remains in effect until the court is able to schedule a hearing for a preliminary injunction. During the preliminary injunction hearing, both parties present their evidence to the court. A preliminary injunction is ordered if it is determined that the plaintiff is likely to prevail in the lawsuit, and the plaintiff is likely to suffer irreparable damage if the injunction is not ordered. A preliminary injunction remains in effect until a final injunction or judgment is issued. The patent owner may not obtain a court order to halt the government infringement. The district court decision may be appealed to the Court of Appeals for the Federal Circuit. Any decision handed by the Court of Appeals may then be appealed to the Supreme Court by writ of certiorari.

Stopping Infringement

Although federal district courts have the final say in issues of infringement, alternative methods may be effective in stopping infringement. Resolving a case of infringement does not have to result in a legal suit. According to the American Intellectual Property Law Association, the estimated median cost of patent litigation is $280,000 up to the trial and $518,000 through trial. Because of the expense of litigation, many patent owners prefer to resolve issues of patent infringement without going to trial. A patent owner may engage in any of the following methods in an effort to stop an infringement:

- File an infringement action in the appropriate federal district court
- Engage an infringement action under a special procedure
- Issue a cease and desist notice to the infringing party
- Negotiate a settlement with the infringing party

File an infringement action in the appropriate federal district court

Filing an infringement action in federal court is the most effective and final method of stopping infringement, but it is also the more costly and time-consuming method. The federal court process involves five phases:

1. **Temporary Relief.** During the temporary relief phase of the process, a patent owner attempts to obtain a temporary court order that restrains the infringing party from performing some act, pending further litigation.

2. **Discovery.** During the discovery phase, both parties to the suit issue one another interrogatories, which are requests for admissions, documents, and depositions of prospective witnesses. Interrogatories are formal or written questions to a witness, which require an answer declared under oath. Depositions are interviews to give testimony or evidence under oath. Discovery is an expensive phase of the process and is often responsible for causing parties to come to some sort of settlement. A discovery also may request confidential information about the patented or infringing invention. The parties are allowed to engage in an

agreement to protect and limit the disclosure of such information. Furthermore, the court may issue a protective order that prohibits public disclosure of such confidential information.

3. **Summary**. During the summary phase, both parties try to obtain a decision as to some key issue, or the patent owner tries to obtain an injunction against an infringing activity. Also, during this phase, the parties may be court mandated to arbitration, or they may voluntarily submit to arbitration.

4. **Trial**. During the trial phase, the matter goes to trial before a jury or judge. However, a Supreme Court ruling dictates that a jury may not interpret patent claims of an invention. Only the judge may make such an interpretation. The jury may, however, make a determination as to whether infringement has occurred and the amount of damages to be awarded. Expert witnesses and visual aides often are introduced to assist the judge or jury in interpreting the technical terms and scientific language that may characterize an invention. Each party must identify its expert witnesses before the trial by filing an Identification of Expert Witness document with the court. Visual aids, such as enlarged copies of claims, may be used to simplify procedures. Charts may be used to compare claims to the invention or to illustrate a sequence of steps. After a verdict is issued, the judge must confirm the verdict before a judgment can be enforced. In some instances, the judge may set aside a jury's verdict if the judge determines that laws and facts do not properly support the verdict.

5. **Appeal**. During the appeal phase, an unsatisfied party appeals the decision of the district court to a higher court, the U.S. Court of Appeals for the Federal Circuit (CAFC).

The CAFC is in Washington D.C., but often travels around the country to hear appeals. The CAFC is a three-member panel of judges that reviews the trial records and determines if a legal error has occurred. If either of the parties is not satisfied with the decision of the CAFC, an appeal may be filed with the U.S. Supreme Court. The Supreme Court rarely hears cases of patent infringement; as such, the CAFC ruling is usually the final ruling on infringement. In some instances, the decision of the CAFC may lead to a new trial.

Expending resources in the initial phases of the process may lead to a successful outcome of winning a temporary relief order against the infringing party or winning an injunction during the summary phase. These outcomes may be adequate to end the issue of infringement before it goes to trial.

Engage an infringement action under a special procedure

Special procedures that are less costly and time consuming than the federal court process have been established to handle infringement issues. Federal and state laws provide criminal penalties for some types of infringement. These laws are not specifically designed to protect patents, but patent owners who have copyrights, trade secrets, and trademarks attached to their inventions may be able to make use of these laws to avoid an infringing action. Federal anti-piracy laws provide for punishment of copyright infringement when the infringement offers a commercial advantage or provides financial gain.

Federal laws also impose punishment for the reproduction of recordings or trafficking of counterfeit recordings, software, and motion pictures. If an infringing party attempts to import copyrighted or trademarked merchandise, the copyright owner or trademark owner may register the copyright, company identifier, or mark with the U.S. Customs Service and have the merchandise seized. If such a seizure takes place, the owner of the seized merchandise will have to appear in court where it will be determined whether the goods should be released to the original addressee or destroyed.

States have statutes that guard against the misappropriation of trade secrets and industrial espionage.

If the infringer attempts to import counterfeit merchandise into the United States, the patent owner may request an order of the International Trade Commission, which will ban the importation.

The ITC will examine the validity of the patent and decide whether the infringing importation has an anticompetitive impact on U.S. commerce. ITC proceedings are less expensive and much quicker than those used in a court of law.

Issue a cease and desist notice to the infringing party

A patent owner may have an attorney send the infringing party a cease and desist letter, which demands an immediate halt to infringing acts, demands an accounting of all illegal sales, and requests payment to reimburse losses due to the past infringement. A cease and desist letter has a well-defined legal meaning because it has to be issued by a judge or government authority, and the attorney would execute the issuance on your behalf.

The letter may offer a threat of litigation if the infringing activity is not halted. In response, the infringing party may do any of the following:

- Ignore the letter.
- Make contact to seek an amicable solution.
- Defend his or her actions in court.

Ignoring the letter is the most frequent action taken in such situations. The patent owner would then have to resort to another method of stopping the infringement. If the infringer makes contact to seek a solution, the solution is likely to entail reneging on the payment of damages in return for a halt of infringing activities. If the infringer decides to defend his or her actions in court, the infringer will institute a declaratory relief action. A **declaratory**

relief action is court judgment that determines the rights of parties without ordering anything be done or awarding damages. The issuance of a cease and desist letter provides the infringer with reasonable cause to believe that he or she will be sued. An infringer may initiate a declaratory relief lawsuit if the infringer has reasonable belief that an infringement lawsuit will be filed. **Declaratory relief** is a request to determine the validity of a patent and determine whether the patent has actually been infringed.

The infringer becomes the plaintiff in the lawsuit and insists that he or she has not committed an infringing act or insists that the issued IP right is invalid. This method of stopping infringement may become costly because the infringer is likely to file action in his or her district, and the patent owner will have to bear the expense of defending the suit in that particular district.

Negotiate a settlement with the infringing party

Negotiating a compromise with the infringing party is the most sensible alternative to resolving infringement issues in most situations. An offer to negotiate a settlement is not considered a threat of litigation, and the infringing party may not initiate a lawsuit for declaratory judgment because of receipt of such an offer. A settlement offer may also provide the patent owner with less money than would be demanded in a court judgment. The settlement may include any of the following:

- Reneging on the payment of damages in exchange for a halt of the infringing activity
- Granting a narrow nonexclusive license to the infringer party so long as the infringement does not involve a commercial identifier
- A reverse license, which gives the infringing party permission to continue manufacture and sales as long as royalty payments are made for all past and future sales

In most cases of infringement or alleged infringement, neither the patent owner nor the accused party wants to engage in expensive litigation. A negotiated settlement offers many advantages that save both parties the expense of litigation and the time necessary to complete the litigation process. Litigation may continue for two or three years when the courts are involved. Furthermore, the payout for a settlement is a guaranteed payment, which does not require enforcement and collection.

A settlement requires that a contract be signed and executed between the parties involved. Some states have specific requirements regarding settlement agreements and contracts. Specific language must be included in the contract to protect the rights of all parties. In some instances, the terms of a settlement agreement must be presented in what is termed a stipulated judgment. A stipulated judgment is a document that must be filed with the courts.

A settlement offer may be reached in privacy through informal procedures known as mediation or arbitration. Mediation requires that the disputing parties submit their arguments to an impartial mediator who assists in reaching a settlement. Arbitration is sought if mediation fails. Arbitration involves referring the dispute to one or more impartial individuals. Arbitration may be initiated by an agreement or by submission from the parties of the dispute. The American Arbitration Association establishes Patent Arbitration Rules and has established a panel of arbitrators. The decision reached during arbitration is usually the final and binding decision.

The International Chamber of Commerce in Stockholm or the London Court of Arbitration governs international arbitration. Companies exist that provide patent enforcement litigation services to patent owners. In return for an agreed upon annual premium, these companies will reimburse part or all of the cost of litigation, dependent upon the limits established by the insurance policy. One company, Patent Enforcement Fund, will provide patent enforcement litigation services in return for a partial interest in the patent. Such insurance coverage may be initiated during the patent application pendency period. Likewise, businesses may be insured against patent infringement through their business liability insurance.

When a party is sued for infringing a patented invention, that party should assert a defense if he or she believes it is applicable. A defense asserted without merit or based on falsehoods may destroy the defendant's credibility and lead to sanctions or even imprisonment if felony perjury is charged. The defendant in a patent lawsuit is most likely to argue that either an established patent is invalid or that his or her invention does not infringe.

Some typical defenses to support such arguments include the following:

- **Invalidity**. The invalidity of a patent is usually based on a lack of novelty or nonobviousness. The PTO may be requested to reexamine any patent that is already in force. The criterion PTO uses in granting the patent is reexamined in an attempt to show that the PTO was incorrect in granting the patent. The defendant will attempt to show prior art exists that renders the patent obvious, or the defendant will attempt to show that the invention was offered for sale, sold, or otherwise disclosed more than one year before the patent application was filed. After reexamination, the PTO issues either a certificate of patentability or a certificate of unpatentability.

- **Inequitable conduct**. In establishing inequitable conduct, the defendant attempts to establish that a patent owner intentionally misled the PTO or withheld material information that would have affected the patent examination process.

- **Patent misuse**. A defendant may argue that a patent owner misused the rights afforded by a patent and, therefore, cannot sue for infringement. Common types of misuse include involvement in unethical business practices or violations of antitrust laws. A tie-in is also considered a patent misuse. A *tie-in* the illegal act of obligating a licensee to purchase another product beyond the scope of the licensor's IP rights. A patent owner who engages in a tie-in may not sue for infringement. The Patent Misuse Amendments Act of 1988 requires that courts apply a "rule of reason" and view all relevant factors to determine if a tie-in arrangement is in any way justified.

- **Lack of standing**. It may be argued that the patent owner lacks the legal capacity, also termed standing, to bring the lawsuit. In such a case, the defendant would need to prove that he or she, not the plaintiff, is the true owner of the patented invention.

- **File wrapper estoppel**. The official file of a patent that is held at the PTO is known as a file wrapper. A file wrapper contains all correspondence, statements, admissions, and documents related to a patented invention. When a patent applicant makes certain admissions or disclaims certain rights relative to the invention, they become part of the file wrapper. Another party may create a similar invention by exploiting such rights and admissions and then design around the patented invention. The patent owner may not sue for infringement over rights disclaimed in the file wrapper. So long as the invention does not infringe under the doctrine of equivalents, the invention does not constitute an infringement. This defense is known as file wrapper estoppel. Estoppel means prevented from contradicting a former statement or action. This is similar to a defense known as a reverse doctrine of equivalents or negative doctrine of equivalents. Under this defense, a device that would otherwise constitute literal infringement is excused because the infringing device has a different function or produces a different result from the patented invention.

- **Exhaustion Doctrine**. Rights to a patented invention are exhausted after the item is sold. The item may then be resold at the will of the purchaser and no infringement occurs. This defense is also known as the "First Sale Doctrine." The doctrine does not apply in situations where an infringing invention is sold without authorization from the patent owner and then resold.

- **Experimental use**. Patent laws allow for the use of a patented invention if the use encourages competition and speeds the release of human health care and certain animal products. The otherwise infringement is allowed when the use of the patented invention is necessary to obtain regulatory approval. The experimental use defense is employed when a patented invention is necessary for research or development of drugs to cure a disease.

- **Repair doctrine**. The repair doctrine dictates that it is not infringement to repair a patented invention or replace components of a patented invention. The doctrine further dictates that it is not infringement, contributory or otherwise, to sell materials to be used for such repair or replacement. The doctrine does not apply to the complete rebuilding of an unauthorized invention or items made or sold without authorization from the patent owner. A specially designed ergonomic chair, for example, may be patented, but the fabric used for the chair is not patented. The repair doctrine provides for the sale of fabric to purchasers of the patented chair. If another company made and sold an infringing version of the chair, any fabric sold to replace or repair the infringing chair would be considered contributory infringement.

- **Claim methods**. In 1999, patent laws were amended to provide a defense that is only applicable to patents that claim methods of accomplishing a process, also termed method claim patents. If a process is invented and a defendant engages in the commercial use of a product produced by the method at least one year before the filing of a method claim patent, the patent may be considered invalid as a defense against infringement.

- **Laches**. Laches is waiting an unreasonable amount of time to file a lawsuit. No statute of limitation exits for filing an infringement lawsuit. However, monetary damages only can be recovered for infringement that occurred in the six years before filing a lawsuit. Despite the lack of statute, most courts will not permit a patent owner to sue for infringement if the patent owner laches. Most courts adopt the six-year period as a reasonable time within which to file suit. Generally, if a patent owner laches for more than six years, courts will dismiss the lawsuit unless the patent owner is able to provide a reasonable basis for the delay.

How to Avoid Scams

Many invention-marketing firms advertise on the Internet, in magazines, on the radio, and late-night TV. Not all of them are running a legitimate enterprise. When a marketing firm is running a scam, they convince the inventor they can guarantee success and use this to justify high upfront fees in exchange for their services. According to Lawfirms.com, a website created to help consumers find lawyers and law advice, statistics show that "such firms do very little, if anything, to actually market the products of investors and boast a success rate of less than 1 percent overall. To avoid marketing scams, make sure to use only the services of reputable industry professionals, and be weary of businesses that promise you huge result for a low monthly fee."

Invention Marketing Scams

The upfront fee many of these firms charge is for an initial assessment of your invention. For the full suite of service, inventors wind up paying anywhere between $2,000 and $10,000. They overinflate your prospects of success and then ask for thousands of dollars to take your invention to the next level. According to the American Society of Inventors (**www.asoi.org**), invention marketing scams cost inventors $300 million a year. In April 2006, a United States District Court judge ordered Davison & Associates to pay back $26 million to inventors ripped-off by what the court called "blatant, varied, and repeated misrepresentations." Less than 1 percent of its customers received royalties that exceeded the fees the firm charged for its services.

If you find the name of a marketing firm, type the name of the company and the word "scam" into a search engine. If the company has a reputation as a scam, chances are that someone filed a report on them, and the report will appear in the search results. If nothing appears, a website called Ripoffreport.com (**www.ripoffreport.com**) allows consumers to file claims free of charge against any business and keeps the report available online for public scrutiny. Unlike the Better Business Bureau, Ripoffreport.com keeps the reports public even if the claimant has been compensated for damages. The public record serves as a business history and prevents the offenders from hiding their intentions from the next customer. The trend of many illegitimate marketing firms is to change their name repeatedly in order to distance themselves from the claims filed online. In 2007, for example, three scam artists barred from misrepresenting their services in a 1998 court order revived the scam under a new name, The Patent & Trademark Institute, and proceeded to charge clients $5,000 to $45,000 for false promises involving commercial licenses and substantial royalties.

Marketing firms found through advertisements are typically scams because most experts in the field of innovation and invention are harder to find and generally do not advertise. Instead, they belong to trade associations and obtain a substantial amount of business through referrals. Unlike scam artists, legitimate firms will not spend their time giving a sales pitch to inventors, but rather, they will give an honest assessment of the invention and suggestions on how to improve it, and they will provide a list of satisfied clients. A legitimate firm will offer an evaluation for a couple hundred dollars, whereas the scam artists charge exorbitant upfront fees of more than $1,000 and will not furnish a list of clients.

Firms convicted of misrepresentations have included false claims about how selective they were regarding the inventions to promote, the objectivity of their market analyses, the percentage of clients whose inventions became profitable, their partnerships with manufacturers, and the source of their income. These firms claim their income comes from sharing royalties with successful clients. In actuality, their profits come from the exorbitant fees they charged inventors for performing almost no meaningful work.

The Spending Spiral

The scammer's entire goal is to lock the inventor into a spending spiral. If the firm advertises itself as an invention promotion or marketing firm and is not part of a university, government agency, or a nonprofit, it is likely to be a scam. There are professionals who can help inventors at various stages of the commercialization process, but few legitimate invention marketing or promotion firms take products from the concept stage to manufacturing or licensing. Illegitimate invention marketers use a variety of methods to lock the inventor into a spending spiral, such as:

- Providing an invention evaluation report that encourages the inventor to pursue commercialization of a product, regardless of its actual merit, to charge further fees for additional, worthless services

- Conducting incomplete patent searches and not rendering any legal opinion about patentability

- Charging large fees to submit a simple and inexpensive provisional patent application, representing it as a regular patent application

- Selling marketing services based on nonexistent connections with manufacturers and other prospective licensees

- Charging high fees to deliver non-customized lists of manufacturers that anyone can obtain

Scammers will goad people into becoming clients by claiming they have contacts at manufacturing firms or that they represent firms looking for inventors. They reinforce the mirage by producing a list of Standard Industrial Classification (SIC) codes that are available free on government websites. SIC codes are part of the government system for classifying industries by a four-digit code. The scammer can look for the type of manufacturing firm you are seeking by entering SIC codes that identify categories of manufacturers, and then pass them off as contacts that match your product. The first service an invention promotion firm will offer is an "assessment" of the invention's potential. As an inventor, you are also less likely to see the problems you may encounter while commercializing your idea. Scam artists take advantage of the inventor's enthusiasm and skewed perspective. Their glowing analysis of your invention is the bait used to hook inventors.

Once you are hooked by the firm's glowing response to your invention, it will ask for a fee for its services. You will not get much for your money. It will do just enough to string you along and get you to purchase the next service at an even higher cost. The subsequent services it will try to sell you may include a marketing report, a business plan, a list of manufacturers, a patent search, patent application, or a professional sketch that shows your invention. By paying upfront fees for a package of services of unknown value, inventors open themselves up to receiving worthless services. Scammers often sell inventors on reports and plans that have little relevance to the invention and industry in which the inventor is working. These products have no value, but they move the scam along. The reason inventors keep sinking money into scams is they feel they have paid this much already, so in order to get the true value from the service, they better keep spending.

Inadequate patent searches and applications

Conducting a patent search for prior art is one of the many services invention promotion firms offer. Your patent application can be rejected in the event that a patent with similar claims has already been filed, so an extremely broad and thorough search for prior art is required. When a division in the USPTO receives your claim, it searches many sources to verify the uniqueness of your claim, including:

- Its own database of existing United States patents
- Foreign patents
- Any published documents
- Technical books and manuals
- Doctoral dissertations
- Scientific articles and research reports in journals and magazines
- Industry and corporate brochures and reports from fields related to your invention

This is a broad set of materials to search. If a scammer only conducts a cursory search, it has no value to you when you apply for a patent, and the application may be rejected based on insufficient research. There are legitimate firms that conduct thorough patent searches. To increase your odds of obtaining a thorough patent search, farm this service out to a patent agent instead. Many are former USPTO examiners, and others are professionals working for legitimate firms located in the Washington, D.C., area and have physical access to the USPTO's files.

Scammers may offer to handle the patent application process. Instead of filing a regular patent, however, they will charge for the cost of filing a regular patent and save money by filing for the less costly provisional patent. Inventors unaware of their status find themselves ignorant to the fact that they have 12 months to file a regular patent application or the claims will be considered publicly disclosed and unpatentable. Provisional patents can be great tools for inventors at the right stage in the process, but paying too much to obtain one at the wrong time can be disastrous to your commercialization efforts. Paying a large fee for a boilerplate marketing report may not bankrupt you, but imagine paying $10,000 only to realize you cannot patent your idea.

Mail Scams

Scam artists also operate by mail. When a patent is issued, a scam artist will glean the information on the USPTO's website, and send a notice in the form of a postcard informing the inventor that a recent search has indicated a later patent similar to your claims was filed in the same subclass. For a fee of $96, plus an addition service charge of $60, the firm contacting you promises to provide a patent number, abstract, drawings, as well as the address of the inventor and manufacturer that is potentially infringing on your claims. The postcard will have the patent number of the claim, so if you suspect someone is filing identical claims, go to **www.uspto.gov**, click "Patents," "Search," and "Patent Number Search," enter the number of your patent, and click "Referenced By." The results will provide a list of patents that cite yours.

Other solicitations by mail include an offer to publish your patent information in an annual directory, an e-mail scam run by someone posing to be an industrialist looking to license patents for manufacture. If someone offers to issue your patent by offering a certificate for payment of fees, keep in mind that the USPTO does not issue such certificates, nor does it send a bill with an amount due after your patent has been published.

How to Avoid Being Scammed

Before purchasing any invention evaluation or marketing services, keep in mind that marketing companies, under the American Inventors Protection Act of 1999, must disclose specific information regarding past business practices through a mandatory disclosure form. In this form, a marketing firm must disclose the total number of clients it has had in the past five years, as well as the number that received positive evaluations versus negative evaluations. If the numbers appear one-sided, the firm may not be legitimate. The mandatory disclosure form must state the total number of customers who received a net financial profit as a result of the promoter's services, the total number of clients who received a licensing agreement, and the names of the promoter's business affiliates. The items listed in the mandatory disclosure form should provide a reasonable estimate of the effectiveness of the firm's services and its direct impact on clients. Also, check your state's Attorney General's office or the Better Business Bureau to see if the firm has any complaints against it. The Federal Trade Commission offers an Invention Marketing Scam pamphlet through its website at **www.ftc.gov**. Websites dedicated to unmasking invention marking scammers include:

- **InventorEd**: **www.inventored.org/caution/list** (site index containing scam watch lists, news on class action suits, and inventor initiatives)

- **IPWatchdog**™: **www.ipwatchdog.com** (online magazine focusing on patent and innovation news and policy, founded by patent attorney Gene Quinn)

- **USPTO**: **www.uspto.gov/inventors/scam_prevention/complaints/index.jsp** (lists current complaints made against marketing companies)

Before working with any firm, create a checklist to determine the effectiveness of the company you have approached. Your checklist should include the following measures:

- Avoid firms that advertise on TV, radio, in magazines, and those that come up high on search engine results when you search for invention.

- Ask for the total cost of the firm's services.

- Work with firms that specialize in one stage of the invention promotion process (for example, patent searching, patent applications, or market research).

- Ask for and check the firm's references, including clients. If it claims it has a network of manufacturers waiting to license inventor's products, check those firms. The list should be long, and you should be able to select a few clients and firms randomly to call from the list.

- Check the name of any firm you are considering working with against all three of the fraud alert lists provided, and avoid any company on any one of them.

- Talk to fellow inventors. For a state-by-state list of inventor clubs and a list of national inventor associations, see the National Inventor Fraud Center's website at **www.inventorfraud.com/inventorgroups.htm**.

- Read the contract before you sign it, and be sure it lists all the services the firm said it would provide.

- Ask for any guarantees in writing, and ask if the firm will provide written results of its patent and invention assessment work.

- Ask for written answers to the following:

  - The number of invention evaluations conducted and the percentage that were positive. Most legitimate firms provide unqualified positive assessments to about 5 percent or fewer of the inventors who contact them.

  - The percentage of inventions that result in a licensing deal in which income exceeds costs and the percentage of invention ideas in which cost exceeds income. Legitimate firms work with a few select ideas and have a decent, but not suspiciously high, rate of success. If they only take on about 5 percent of inventors who contact them, they should do a good job of helping those they cherry picked. Remember, many firms lie about these figures, have been successfully prosecuted, and paid large fines. Many of these firms are still in business. This is a testament to how much money these scam artists make off gullible inventors. Remain skeptical, and insist on answers in writing.

- ♦ Names of all the invention companies this company or its officers have been affiliated with in the past ten years and other names the company operates under anywhere in the United States.

- ♦ Has the company ever been investigated by the Federal Trade Commission or any state's attorney general's office?

- Ask what industry the firm specializes in. If they claim they help inventors with everything from food products to medical devices, toys, and sports equipment, be skeptical. Not only is the process of commercializing an invention made up of discrete steps in which higher quality firms specialize, but also invention promoters who claim to have contacts with potential licensees must specialize by industry, as there is no realistic way they could effectively keep up with multiple markets (note that consumer products like housewares and giftwares can be lumped into a single category).

Most invention ideas start out with significant shortcomings, which may be addressed as the inventor learns more about the field. If the marketer says the invention is market ready, it sounds too good to be true. Find out how the firm makes its money. If it says its makes most of its money from royalties, request documentation, and ask why it charges fees for marketing reports.

What legitimate services look like

Although many invention evaluation organizations charge a few hundred dollars for an evaluation, legitimate agents or representatives who market inventions to prospective licensees charge royalties only. These two activities are separate functions. Legitimate invention evaluators will charge a fee to offset their overhead costs. Scammers will offer a free assessment. Legitimate marketers, such as agents, will charge a royalty to represent your product while scammers will charge a flat fee to provide you off-the-shelf marketing reports and impersonalized, publicly available lists of contacts at manufacturing firms.

Inventors can get information about inventing processes, pitfalls, and techniques from inventor clubs or associations. For a list of inventor associations in various American cities, see *Everyday Edisons!* at **www.everydayedisons.com** under the *Resources* section. *Inventors Digest* magazine at **www.inventorsdigest.com** and the United Inventors Association (**www.uiausa.com**) are two well-regarded independent inventor resources. Websites with tips on a variety of invention-related products can be found at:

- **Virtual Pet**: **www.virtualpet.com/invention** (provides information and tools to assist inventors commercializing new products)

- **Patent Café**: **www.patentcafe.com** (full service provider for IP solutions, including consulting, training, customization, and patent data feeds)

Invention Promotion Laws

The good news is that the crackdown on this type of fraud is increasing, as perpetrators are being handed significant jail time and ordered to forfeit all assets related to their scheme. Under the American Inventors Protection Act of 1999, otherwise known as the Invention Developer's Law, Section 297(b) states that an invention promoter who causes damages because of false claims in its disclosure is subject to paying statutory damages of up to $5,000, including attorney fees incurred by the client because of litigation. To register a complaint against an invention promoter, the

American Inventors Protection Act of 1999 allows the USPTO to accept filed complaints against invention promoters, forward these complaints to the accused, and make the complaint and response publicly available on the USPTO website. Do not submit original documents as proof of the claim. A complaint filed under the American Inventors Protection Act must provide the name and address of the complainant, the client, and the accused, as well as an explanation of the promotion services that were offered, the media outlets the promoter advertised with, the nature of the business relationship, and the complainant's signature. Should any of the required information be left out of the complaint, the submission will not be processed. Complainants also have the option of withdrawing complaints before official publication with the USPTO. Complaints should be mailed to:

Director, Office of Independent Inventor Programs
Box 24
Washington, D.C., 20231
Phone: (703) 306-5568
Fax: (703) 306-5570
E-mail: **independentinventor@uspto.gov**

If you feel that you have been exploited by an invention promoter and need legal consultation or assistance, Penny Ballou is considered one of the foremost authorities on handling issues that inventors have with invention promoters. She is the founder of an inventor's group and serves as advisory board president of **www.inventored.org**. She is also a former Licensing Executive Society member and serves as a consultant for the Professional Inventor's Alliance at **www.piausa.org**. She has made her contact information publicly available and can be reached by phone at (702) 435-7741.

CASE STUDY:
AVOIDING MISTAKES

Cindy Kroiss, president
Sick Bear Inc.
info@sickbear.com
www.sickbear.com

As a mother of two young daughters, I frequently had problems cleaning up vomit when both of them got sick. In the car, they always got nauseated, but when one of them developed a nasty case of the flu, I came up with an idea for a spill-resistant container children could hold when they needed to vomit.

When I visited a hospital, they gave me a flimsy little kidney-shaped bowl. It was too unstable to put in a moving car and not very comforting for a child. I searched the Internet and found there was no container specifically designed for children to vomit into. Everything currently on the market was designed for adults. My search for a cute, spill-resistant bowl for vomiting became the beginning of my market research for development of the Sick Bear Bowl.

When I spoke with other women in my workplace, I realized I was not alone in constantly having to deal with sick children. The crisis mode kids go into when they are feeling sick is frightening for them, and that is how it gets messy. I learned that children with chronic illnesses such as cancer might experience nausea and vomiting following radiation treatments. I wanted to design a container for them, so they would not have to use a kidney dish.

The Sick Bear Bowl was designed with two big handles that make it easy for children to hold and is decorated in pastel colors with the Sick Bear trademark. Three weeks after I began market research, I sketched the first prototype. Three months later, I founded my company, Sick Bear, Inc.

I used money from personal savings, family loans, and zero-interest credit cards to finance the business. I was surprised at the amount of financial investment required to bring the Sick Bear Bowl to market. You think you are financially prepared, but after patents, trademarks, product development, business startup costs, marketing collateral expenses, launching a website, and your initial order, you realize the investment is much more than you could have ever imagined.

I never considered leasing or selling my idea and wanted the challenge of manufacturing my Sick Bear Bowl by overseeing the production of the CAD drawings, the prototypes, the injection mold, the samples, and fixing any flaws. I designed it specifically to meet a need for my own children, but I am glad it has been helping others as well.

Although I was surprised by the extent of personal investment required, my mistakes did not impede me. In retrospect, I probably should have ordered a smaller number of Sick Bear Bowls manufactured for my initial order because it is a big chunk of money, and storage space became an issue I was not ready to handle. My advice to first-time inventors would be to use their existing skills and talents to learn the business end of being an inventor. Being prepared for the challenges of running your own company can save you thousands of dollars when marketing your product. I used my digital arts skills and designed my business cards, brochures, and trade show banners. I also refreshed my HTML skills so I could design and launch the Sick Bear, Inc., website.

Coordinating Your Commercialization Strategy

When it comes to commercializing a consumer product, no two strategies are alike. In fact, what may work in some markets may fail in others. Preliminary commercialization begins with market research, but coordinating a strategy means fashioning that research data into action. Putting data into action means narrowing down your list of manufacturers to one, choosing between licensing offers, deciding what patent claims should be most protected, deciding which terms are negotiable, and knowing what market segments you will go after. If you have gotten this far with your invention, you have:

1. Documented and refined your invention
2. Developed a preliminary strategy outline
3. Evaluated its commercialization potential
4. Researched your product and industry
5. Identified business partners
6. Filed patent applications
7. Avoided invention marketing scams

After this point, you know everything about the product — its design, its benefits, which industry it serves, how it should be distributed, its competition, and so forth. You are ready to promote your invention either by licensing it or building a business around it. Having a patent registered by the time you are ready to start selling is important to the goal of maximizing potential profits. So, what else do you need beforehand? Much like a driver who is ready to put a new car on the road, you will need insurance for your patent in the event that someone tries to infringe on your product and damage your profits.

Types of Insurance

For a fee of approximately $1,500 a year, The Intellectual Property Insurance Services Corporation (IPISC) will cover any litigation expenses on claims up to the amount insured, roughly between $100,000 and $1,000,000. They also will conduct their own inquiry into your claim to determine whether there is a legitimate case or not. If a potential infringer sees you have a legal team conducting an inquiry, it may cause the infringer to negotiate a settlement and pay you royalty fees. Consider getting patent insurance only if your product's potential royalty fees are substantial enough, as the cost of the policy can be expensive. Most patent lawsuits deal with a lot of gray area when it comes to claims.

A patent attorney may be necessary if a licensee includes an indemnification clause in the terms of their licensing agreement by which they are exempt of responsibility for events that may occur beyond their control. For example, if a distributor has a fire in their warehouse and your product is destroyed, the manufacturer may not want to be held responsible. In such an instance, the manufacturer may expect the inventor to handle any expenses involving litigation against a distributor. This may be an issue of quality control, so if you have been allowed some supervision over quality control, or feel comfortable with the level of quality control at all levels in your distribution tier, it may be acceptable to allow for such an indemnification. As a rule, the only provision you should agree to indemnify is when a third party sues the licensee or licensor, which can occur when a product injures a consumer.

Liability insurance may be needed if your invention injures a consumer, and the licensee claims it cannot be held responsible for manufacturing to specification. Keep in mind that although most inventors carry it, liability insurance can be difficult for independent inventors to obtain. According to Alan Tratner, founder of Inventors Workshop International, many insurance brokers are not willing to deal with the risks associated with entrepreneurial product innovations. Many brokers do not like wasting time quoting prices to inventors if they are not backed by a distribution tier or funded by a university or some other business organization because of the statistical likelihood that their product will never make it to market without any type of backing. When looking for a broker, approach other small business owners, and ask for referrals instead of consulting the yellow pages. Note that the difference between an agent and a broker is that a broker represents individual clients, whereas an agent works with companies. After finding the right broker, describe the invention to him or her. If the invention is highly specialized, ask the broker if any insurance company specializes in providing directly related coverage to your field. Price quotes can be lower depending on how much the insurer knows about your product, its industry, and the known risks involved. When visiting a broker, you will be asked questions that will serve as the basis for the risk evaluation, such as:

- Who is your manufacturer?
- What does your product claim to do?
- Has your product been tested by an expert?

Types of Royalties

In gearing up for commercialization, make sure you will be receiving a monthly or annual sales report and that the royalty rate is clearly defined. Expenses not associated with production costs, such as returned units, rebates, and sales commissions, should be deducted from the sales reports you receive from the licensee because they will impact the sales measure by which your rate is paid. Most marketing experts believe that promotion and advertising deductions should not be included in net sales because that is not a licensor expense. Make sure the licensee knows that you do not expect these deductions in net sales when calculating royalty rates. Also, know how long it will take the product to be made and shipped because the longer the time to market, the larger your upfront payment should be. You also can induce the licensee to pay royalties quicker and in a timely manner by establishing late penalties and high interest rates for overdue payments, which companies will want to avoid. Monthly rates are commonly set between 1 percent and the maximum rate permitted by federal law. Keep in mind that interest rates allowed by the Federal Reserve are adjusted periodically; therefore, current rates are subject to change. To find out the current interest rate, go to **www.federalreserve.gov**, or contact a local Federal Reserve District Branch to verify the current rate. There are 12 branches, located in Boston, New York, Philadelphia, Cleveland, Richmond, Atlanta, Chicago, St. Louis, Minneapolis, Kansas City, Dallas, and San Francisco. If you license a product in which the price is sure to drop over time, request a royalty rate based on per unit sale rather than net sales because the price drop will have an adverse effect on net sales, even if the same number of units are being sold. Technology products such as computers tend to drop in price. Again, market research of similar products should indicate the direction of the trend. If the product is considered a method or process, base royalties on the number of times the method or process is used to do something. Keep in mind that royalty rates may fluctuate based on the number of sales or inflation. If sales reach a certain milestone, the royalty rate agreed upon should increase.

A different type of royalty occurs when a hybrid license is created. A **hybrid license** is defined by antitrust laws as a license that assigns rights from a patent issued or pending, as well as an additional form of intellectual property, such as a trade secret or how-to. A hybrid license protects the licensee from having to pay the same agreed upon royalty rate in the event that your patent application is rejected. If your idea is not patentable, but the agreement has the company locked into an agreement with you, it still may have to honor its contract but may change the terms of the royalty rate. Some inventors fear receiving no royalties as a result, and therefore, structure a guarantee for a minimum annual royalty rate, which ensures that they receive a royalty regardless of any sales underperformance. If a third party asks to incorporate its trademark into the sale of your product, royalties will have to be split with the third party because the presence of their trademark presumably would be increasing your net sales. If the manufacturer is producing your product at an offshore facility for domestic distribution, your royalty rate (termed F.O.B. for "free on board") should be adjusted about four points higher than the net sales rate.

At some point, you will have to license your business in your state and possibly, in your city as well and pay business taxes on the revenues you generate. Your Secretary of State's office and state and local departments of revenue can provide information about timelines for incorporation and tax purposes. If your product is taxed locally, a manufacturer will make a deduction on net sales. If the manufacturer is making a lot of deductions against net sales that include freight, credit, taxes, returns, and discounts at the time of sale, make sure it does not make deductions for sales commissions, debts, promotions, marketing, and advertising. If it balks, ask for a provision in the agreement that guarantees the total percentage of deductions never exceeds more than 10 percent of gross sales.

Marketing and Publicity

At some point before production or immediately after production of the invention has begun, the inventor will need to publicize the invention. Several media are available to the inventor to publicize his or her own invention, but the inventor must be diligent in finding the type of publication that prints articles on products aimed at your targeted end user. The inventor may engage in media publicity, public exhibitions, premium marketing, celebrity endorsements, or other marketing tactics to promote the invention. Other publicity ideas may be found in books, journals, trade magazines, periodicals, and other references or research publications in the field of the invention.

Niche marketing

The general belief about niche marketing is that the more narrowly it can be identified and defined, the more space within that niche a company can control. More and more companies are gearing products toward a specific consumer. Sweet, carbonated alcoholic beverages for the college and young adult market would be an example of a narrowly defined niche. Within this niche, companies test market for design. For example, test markets may indicate that low-income college students who drink sweet carbonated beverages may prefer to drink these beverages in single cans, whereas students from higher-income families may prefer bottles. Niches are another way for companies to connect to valuable customer groups. If your evaluation indicates that your idea is inherently appealing to a narrowly defined niche demographic, you are more likely to find a licensee willing to become a partner in your enterprise.

Price

Price is another method for targeting a desired demographic or consumer niche. Cheap products are intended for lower-income buyers. Those that are more expensive to make are geared toward consumers who can afford their price. Discounts, rebates, and coupons are techniques for using price to appeal to a specific type of customer. How will price affect your invention's target customers' buying decisions? Successful businesses use these techniques because they work.

Marketing to catalogs

Catalog companies are looking for novel products that are not available elsewhere. The niche of catalogs is it allows them to introduce unique products at lower prices. Grey House Publishing produces the Directory of Mail Order Catalogs, which lists 9,000 catalog companies and includes information on the product lines they carry, as well as contact information for their buyers. This is an expensive book and is likely available at larger libraries' reference desks. The National Mail Order Association's National Directory of Catalogs is another resource that can guide inventors in their search for catalog buyer contact information. Catalogs.com offers catalog houses listed by category, so find the categories that best fit your invention, call their customer service number, and ask for a list of their buyers. Mail your sell sheet and marketing package to as many catalogs as possible, and match your product with catalogs based on their market position.

Marketing position is important because some manufacturers position themselves as low cost leaders, while others strive to be known for luxury, reliability, or providing products on the cutting edge. Catalogs position themselves with niche offerings, so select catalogs whose niche offerings are complemented or extended by your invention. Determine their market position by going online and reading the company's online catalog. For example, the Horchow catalog company summarizes its market position by stating on its website: "Our name means more than home decor. Our catalogs carry the finer things in life: classic, elegant, and beautiful products. Products your home deserves."

Catalog Reps

If this all sounds too time-consuming, consider using a catalog rep. The National Mail Order Association (**www.nmoa.org**) offers a product rep program. For a 7-percent commission and a $500 membership fee, the association will contact medium-sized catalog companies on your behalf. Catalog Solutions (**www.catalog-solutions.com**) is a catalog rep company that claims to have placed clients' products in more than 200 catalogs. Remember to check out any organization that claims to help inventors using resources listed in this book.

Media promotion

Certain media outlets provide free publicity for new and interesting inventions. Radio shows, TV shows, newspapers, and magazines offer free publicity. Local radio and TV shows seek interesting guests and some may even offer inventors an opportunity to demonstrate or discuss their inventions. An inventor would need to seek out appropriate shows and make contact with the producer by sending a press kit or letter indicating how and why the invention would be of interest to the listening audience.

Many magazines feature new inventions and ideas; some may also include a feature article of the new invention or idea if the invention holds the interest of the editor. Magazines that feature new inventions and ideas include:

- *Apartment Life*
- *Argosy*
- *Better Homes and Gardens*
- *Changing Times*
- *House and Garden*
- *House Beautiful*
- *Jet*
- *McCall's*
- *Mechanics Illustrated*
- *Outdoor Life*
- *Outdoor Living*
- *Parade*
- *Playboy*
- *Popular Science*
- *Sunset*
- *This Week*
- *True Story*
- *Advertising Age*

Invention contests

For inventors who are not concerned about protecting trade secrets and are simply looking for funding and recognition, invention contests are available on television in the form of reality shows. *Everyday Edisons!*, for example, is a PBS show that documents the invention commercialization process. Although the odds of being selected from open casting calls is low, inventors selected receive free consultation and investment in product development. Find out more at **www.everydayedisons.com**. Another reality show called *American Inventor* brings useful consumer products to market. Inventors can submit up to five different ideas, and a panel of producers screens those selected in a two-minute presentation. To find out more go to **www.fremantlemedia.com/home.aspx**. Additional invention shows and contests include:

- "Mothers of Invention," sponsored by Whirlpool and Mom Inventors, Inc. (offering a $5,000 grand prize and 5 percent royalty)

- "Invention Quest," sponsored by Staples, pays $25,000 to the grand prize winner along with placement in Staples stores.

- "Search for Invention," sponsored by The Hammacher Schlemmer catalog

- "Startup Demos," hosted by The Massachusetts Institute of Technology Alumni Association's Enterprise Forum

- "The Technology, Entertainment, and Design Conference," a membership-based event highlighting computer innovation

- *Modern Marvels* on The History Channel hosts an inventor's contest in conjunction with the National Inventors' Hall of Fame. The grand prize is $25,000 and a meeting with an invention consultant.

Exhibitions

Exhibits, trade shows, and business shows offer an opportunity to publicize an invention and provide a hands-on demonstration of an invention's techniques and capabilities. Thousands of these types of exhibitions are held throughout the country each year.

The inventor must seek out those exhibitions appropriate for the type of invention being presented and arrange to present the invention attractively. These types of exhibitions and showings are relatively inexpensive, in the range of a couple hundred dollars. In most cases, the inventor is provided a booth or space to showcase the invention. A working model of the invention should be displayed, so it is attractive and interesting to the audience. The inventor should also use the opportunity to provide the audience with literature that promotes the invention. One of the disadvantages of this type of promotion is that others will have an opportunity to copy the invention in hopes of selling the invention or avoiding its patents. Sponsorship at a sporting event or other public function is also a relatively inexpensive method of showcasing an invention and acquiring publicity.

Promotion

Promotion refers to the way products are marketed to their target consumer. For example, when companies promote their products, they associate their product with seemingly unrelated images to evoke certain emotions. Although an alcoholic beverage may have no immediate association to the image of an attractive woman, a clever advertisement will attempt to form a connection between the image and the use of their beverage. In this particular case, if the advertisement is targeting men who drink beer, this image of attractive women may be used to grab their customers' attention by associating their brand with deeper desires (sexual) or the notion of having a good time. Different promotional strategies are used to promote to different audiences. For example, a promotional strategy geared toward parents who consume alcoholic beverages would not involve images of women but more likely images associated with holidays or vacations. How will your product appeal to its target audience? What sort of creative promotion would capture consumers' imaginations? How would the advertising message be delivered? What would its packaging look like? How much might it cost to advertise enough to generate business for your invention?

The Sell Sheet

A sell sheet is a one-page flier that gives technical information related to your product. The purpose of the sell sheet is to communicate that your invention works and is marketable. Sell sheets can be used to obtain orders if you are pursuing a strategy that involves test marketing. To create a sell sheet, start by devising a layout. Do not go overboard with graphics; include a logo with room for a headline, a larger intro paragraph, and about two paragraphs of selling text that summarizes the purpose or function of the invention, pricing information, and its benefit to end users. You will need an illustration of the finished product or photographs of a "fit and finish" prototype. Leave enough room in the lower right corner of the sell sheet for your representative's contact information or the information of whoever has been assigned to process orders for a run.

Printers that assist clients in layout can help design the sell sheet. Surf the Internet for printers or art design companies that specialize in sell-sheet layout and brochures, and have your chosen company produce a small run of sell sheets or create an electronic file. Before you contact customers, you should have something to show them, such as a 3-D drawing or specifications for the manufacturer. A proof-of-function prototype is also a low-cost way to advance your marketing effort. You may find this "proof of function" prototype is sufficient for a prospective licensee to see the value of your invention. This is most likely to be true if you are pitching a manufacturing firm used to seeing rough versions of products and would want to finalize the design themselves.

If you do sell to a catalog house, you will need to be prepared to deliver a large volume of manufactured products to them.

The Price List

In addition to a sell sheet and a prototype, your marketing materials should include a price sheet that lists the price per item at different run sizes. A "run size" qualifies as the number of units the buyer can order at specific prices set by specific volume. Catalogs deal with inventors because they want to offer their customers new products at great prices and profit from a large markup margin. Price your product in accordance with what catalogs normally expect to pay for product offerings similar to yours. With large catalog companies, you will be manufacturing in high volume, so offer them the best price possible based on volume. If a catalog company offers to take the product on a trial basis, consider manufacturing a small test run at large volume pricing. Doing so gives the catalog company an incentive to try your product for a test run. If you let them know you are making a special offer, they may appreciate your gesture and order the highest volume because of the value your special offer presents.

Terms and Conditions

When soliciting catalog houses or manufacturers, it is customary to draft a one- or two-page sheet that lays out your business terms and conditions. This sheet should cover items such as:

- The period for which prices listed on your price sheet are applicable
- How order cancellations will be handled
- How return of faulty or damaged goods will be handled
- Delivery methods, timeframes, and shipping charges
- Customer recourse if the inventor fails to deliver the product on time

- Payment terms (catalogs usually pay after receiving the product; you can require they pay within 30 days and offer them a discount of 2 or 3 percent if they pay within ten days of receipt of your invoice)

- Termination of relationship (your right to discontinue an agreement if the customer fails to pay)

- Minimum order information

- Your ability to customize the product for a catalog

Other Marketing Materials

You may wish to provide possible advertising copy to customers by including a description of features and benefits, along with taglines or mock-ups of catalog entry blurbs. This material should not go on a sell sheet, which works best when it remains streamlined and generic. You do not need to be a copywriting expert to write a sample blurb about your product. Read product entries in a few of the catalogs you are planning to submit to and draft some language consistent with what you see there.

You may want to consider providing the catalog buyer with a toll-free number they can call if they have questions about your product. Although catalog houses are used to dealing with independent inventors, the more customer service and marketing savvy you demonstrate, the better your relationships will be. Some catalogs will want more photos than what is available on your sell sheet. They may want photos of the product in use or several photos of the product to choose from. If this is the case, you may need to hire a professional photographer specializing in business photography to create a set of digital photos.

Press kits

Once you have compiled marketing materials, you have all the basic components needed for a press kit. If you can garner free publicity for your product, interested catalog buyers may contact you. Catalog buyers look through consumer magazines' new product sections to identify products they might like to carry. These sections are usually in what is known as the "front of book" section of magazines, where many short articles are carried. Sometimes new products sections are called "roundups" or "new product reviews." Your press kit should include:

- A cover letter addressed to the editor of the new products section by name that briefly talks about your product, what is new and exciting about it, and any personal story that might capture the editor's imagination or provide the basis for an interesting short article
- Your sell sheet
- A fit and finish prototype or product sample packed in prototype packaging materials, if available
- Professional photographs of your product

Vogue calls their new product roundup section "Shopping Alert." *Architectural Digest* has a section called "Great Design for Under $100." *Modern Bride* has an "Essential Accessories" section covering shoes, veils, and purses. *Cookie*, a family magazine, has a "Shopping and Gift Guide" in every issue. *Lucky*, an entire magazine devoted to shopping, has a section called "Editors' Obsessions." If you can match up a magazine to your product, send the editor a sample; you may receive valuable free publicity in return. A trip to a large bookstore or newsstand can help narrow down target audiences that might be interested in your product. Mail your press kit and product sample to the editor of each magazine's new products section. If your product is featured, include copies of these clippings in the marketing package you send to catalog buyers.

CASE STUDY: MARKETING IN THE GARMENT INDUSTRY

Kathleen Fasanella, president
Apparel Technical Services
P.O. Box 12323
Albuquerque, NM 87195-2323
kathleen@fashion-incubator.com
www.fashion-incubator.com
Phone: (505) 877-1713

I have worked in the garment industry as a product design engineer for nearly three decades. I have worked extensively with inventors over the past 15 years, helping them with the technical aspect of engineering, prototyping, feasibility studies, product marketing, manufacturing, and sales. The biggest problem for inventors is feasibility; many have unique ideas but the potential demand is often missing. As a result, it is common in apparel-related products that few parties are interested in licensing an idea. This does not mean the product is not worthwhile; it just means the inventor is going to need a bit of support to establish a market and get the best return for his or her innovation.

Apparel innovators have a broad spectrum of alternatives. The most effective way to market in the apparel industry is to create a Web page with shopping cart, distribute press releases via sites such as PRwire.com, and invest in social media. Blog about your product, and contact bloggers to write about it. This industry requires very little startup, so one of the first things you should do is create a prototype. I suggest test marketing the prototype, and then consider tightening your intellectual property protection because in the sewn product industry, very little is unique. Licensing usually means leasing the rights to use someone's name for a product they already manufacture.

Because the catalog side of the apparel industry is more seasonal and varies accordingly, many inventors I have worked with are surprised at how a single item they have created will need to evolve accordingly. When developing a garment prototype, the inventor needs to consider color as an important aspect of test marketing. In order to test-market, you may need to change the color according to each season; pastels in spring, bright colors are worn in summer, autumn colors in the fall, and gold, white, and silver for holidays. Catalogs will expect goods to follow seasonal guidelines even if the product itself is static.

If you are able to generate a positive response from your test market, the next step is to hire an independent sales representative who will sell the product via their network of buyers — the different types of wholesale shows at which they lease booth space and possibly, niche trunk shows depending on the item. Most products suitable for distribution have a longer shelf life, but garment goods are too seasonal. There is some distribution but it is limited to commodity products such as white tees, tube socks etc. I do not recommend that anyone get into commodity products unless they have a lot of money to fund it. It requires a great deal of sophistication with respect to electronic data interchange and vendor compliance, and the margins are very thin.

Foreign Markets

Congratulations! If you have managed to make it this far in your commitment to bringing an invention to market, it means you have accomplished a lot. You have made a commitment to your idea, researched the market, registered a patent, established a distribution tier, and are earning royalties. Now what?

The world has become a globalized marketplace, offering new opportunities for inventors to find growing markets outside their domestic market space. Selling products to foreign markets requires much of the same due diligence discussed in this book. U.S. patents protect your intellectual property in the United States. To obtain protection elsewhere, foreign patent applications are required. In some cases, foreign patents are not worth the money and effort required to obtain them. Foreign patent applications will require an attorney's assistance.

Foreign Patent Applications

Many countries' governments are signatories to international intellectual property treaties and conventions that govern filing procedures for foreigners. The Paris Convention entitles a patent applicant in any member nation the right to file a corresponding patent application in other member countries within a year of their earliest filing date. Other treaties exist offering the same type of reciprocal rights to member countries.

Obtaining foreign patents is expensive, complicated, and only worth pursuing if a large market exists in the countries you are targeting. Before seeking a foreign patent application, it is advisable to have an offer in place for a foreign licensee agreement, as some of them may be willing to pay for the cost of the application. According to David Pressman, author of, *Patent it Yourself*, "Filing [for a patent] in the United States usually gives you ten to 50 times more bang for your buck than filing abroad, which costs ten to 50 times as much anyway."

Basic foreign prospect requirements

Because of the expenses and complexities involved in prosecuting a foreign patent, it is important to know if your invention's presence in foreign markets is worth the investment. Market research into foreign markets should indicate sales projections that will result in at least:

- $500,000 in sales
- $100,000 in royalties
- One contract with a licensee willing to pay a substantial up-front payment, royalties, and cover the cost of filing a patent for a foreign application

Many experts also agree that it is never a good idea to file a foreign patent application for the sake of recouping damages incurred by infringement overseas. Even if a foreign application is obtained to stop infringement, the market where the infringement is occurring may be smaller than the suggested number of sales recommended to pursue a foreign patent in the first place. In addition, the expenses incurred by litigation, maintenance fees, and filing may not be enough to justify establishing priority there.

Filing with Treaty Organizations

The first step in filing for a foreign patent application is to learn about the various filing rules governed by each of the international treaty organizations. International Convention for the Protection of Industrial Property, otherwise known as the Paris Convention, is an international treaty followed by most industrialized nations in the world and is governed by the World Intellectual Property Organization (**www.wipo.int**). By filing a RPA or PPA in any of the countries that observe the rules of the Paris Convention, you are allowed to file in another country under the same jurisdiction one year from the filing date in the original country you filed in. For example, if you file a PPA or RPA in the U.S., Italy will honor the one-year rule from your application filing because both countries are members of the Paris Convention. Keep in mind that the one-year rule applies for utility patents, whereas design patents are allowed six months. The rules also state that in order to receive the same benefits on a foreign application, you have to claim "priority" on the RPA or PPA you file in the U.S. Under the Paris Convention, if you fail to file a foreign application after one year, foreign countries that observe the Paris Convention will accept a foreign application under the condition that the invention has not been patented or sold yet. However, it will be considered a non-Convention application, and you will lose some of the benefits guaranteed under "priority," such as proof of prior art.

To sell in multiple foreign countries, filing an application under the Patent Cooperation Treaty (PCT) allows you to file one application instead of having to file separate applications for each individual jurisdiction or country. For a list of other PCT and Paris Convention members go to **www.wipo.int/pct/en/texts/pdf/pct_paris_wto.pdf**. It should be noted that the period to apply for patents in additional foreign countries after a PCT application has been filed is 30 months. By filing a PCT, the USPTO will conduct its own patent search to determine your application's patentability within the foreign markets you are trying to penetrate. It is important to educate yourself on PCTs. To learn more about these treaties, consult the USPTO website on foreign applications. Chapter II of the

PCT filing application also grants the USPTO the power to conduct an examination of the prototype to determine which claims will be allowed, rejected, or need to be revised in a report citing references. The World Intellectual Property Organization also governs the PCT. The PCT's fees are published in the *Official Gazette*. With an application, you will be required to pay a transmittal fee, a patent search fee, an international filing fee, and a fee for priority document. To contact the PCT, contact the PCT department of the USPTO (**www.uspto.gov/patents/init_events/pct**); call (571) 272-4300, or mail them at:

Mail Stop PCT
Commissioner for Patents
P.O. Box 1450
Alexandria, VA 22313-1450

A handful of countries, however, are not members of any international treaty. The time period to file in a non-Convention country is six months after any U.S. filing date. Most non-Convention countries are nonindustrial countries, with a few exceptions, such as China and Taiwan. In these countries, you will not be able to establish priority and will not need to produce a copy of the original patent application. It is only required that your invention has not yet been made public, or you have already been granted a foreign filing license with your U.S. filing receipt. In some countries (even those belonging to a convention), a foreign filing license must be granted before filing for a foreign patent application. A foreign filing license gives the inventor permission from the host country to conduct business there. Some members of treaty organizations, such as the United States, Germany, France, the UK, New Zealand, South Korea, and Canada, also require filing licenses. Countries that do not require a foreign filing license regardless of membership are Mexico, Japan, Taiwan, Indonesia, and Australia. Under U.S. law, willful failure to obtain a foreign filing license can carry a fine of up to $10,000 and imprisonment of two years. If the invention involves multiple parties and was created in multiple foreign locations, seek the advice from a patent attorney on how to proceed with a filing.

Filing a PCT covers all 180 participating countries and is even recognized by the European Patent Office (EPO) at **www.epo.org**. The EPO is located in Munchen, Germany, and has a separate facility in the Netherlands. The EPO also honors one patent filing for multiple countries under its membership. Once registered as a European patent, its filing date will be considered the filing date of the original U.S. patent application, establishing priority further back in time. However, patent examination under EPO will be more stringent than PCT and requires a European patent agent to facilitate the process. Conversely, European patent agents are better trained than those who work for the PCT, and their assistance is more likely to define your claims more accurately, which reduces the chance of them being rejected later on. Any application granted a patent with the EPO has a life of 20 years. A list of current EPO members can be found at **www.epo.org/about-us/organisation/member-states.html**.

In addition to providing application assistance, foreign patent agents will know the laws that exist in the country where you want protection. To find a foreign patent agent, ask for a referral from a U.S. patent attorney through one of the previously mentioned methods or through attorneys who advertise in the *Journal of Patent and Trademark Office Society* (**www.jptos.org**). If you cannot find a U.S. patent attorney who can provide a referral, try typing the keywords "patent agent" and (the name of the foreign country) using a search engine. It may be helpful to do an engine search for the website of the foreign country's patent office and contact the foreign patent office for referrals. Most embassies will have this information as well. Some patent attorney firms specialize as intermediaries and handle foreign patent filing exclusively. These intermediaries — usually based in larger cities like New York or Los Angeles — will do the work in finding a qualified patent agent abroad and will work closely with the foreign patent agent on all the details regarding your application. These attorneys in effect will act as your agent to the foreign patent agent. If language barriers are a concern, some British law firms specialize as intermediaries to other European countries because they may speak the language of that country and have a good working knowledge of patent laws in that region. Conduct an initial foreign patent search for prior art before contacting an agent; go to **http://ep.espacenet.com/?locale=en_EP**.

In the U.S., if the RPA or PPA registration receipt does not include a foreign license grant, that means you are not allowed to file a foreign application for six months. Such an occasion is rare and may only apply if your invention has a potential military use. In other words, the six-month period following your application date allows the government to review your application for a possible national security classification. Patents reclassified under national security will not be eligible for foreign applications and will be placed under a Security Order from the USPTO until the patent has been officially declassified.

If the invention is of use to the federal government, the inventor should contact the General Services Administration (GSA) to offer the product. The GSA will respond by sending an inventor the appropriate forms and instructions for submission. If it is an energy-related invention deemed favorable by the National Bureau of Standards, the Department of Energy may provide inventors with a research grant to further the invention. Also, for inventions that may be of use to state and local governments, the inventor should contact the appropriate procurement offices.

Even though you have one year to file a foreign convention patent application after filing a PPA or RPA domestically in the U.S. to establish earlier priority, it is never advisable to wait until the end of the filing period to do so. Foreign agents will need to time to prepare the documents, search for prior art, and translate documents from your U.S. patent registration. Although filing an application too early may be counterintuitive to your commercialization efforts, filing too late puts the application in danger of losing the grant. If you have to put off filing as a result of your commercialization efforts, make sure an application is filed no later than three or four months before the end of the one-year rule.

Foreign Patent Laws

Before the General Agreement on Tariffs and Trade (GATT) was formed, most countries had different patent laws. GATT greatly enhanced the uniformity of patent laws among nations, and the organization lasted until 1993, when the World Trade Organization replaced it. Although GATT no longer exists, the accords are still in place regarding patent laws. Despite the general uniformity, every inventor seeking foreign patents needs to be aware of the remaining differences between U.S. and international patent laws. In the U.S., all patents must be filed in the name of the inventor, but in most foreign countries, the organization assigned the rights, such as a manufacturer, can apply for the patent in the organization's name. Although the U.S. and countries in the EPO conduct evaluations to determine whether the application is novel and nonobvious, smaller countries, such as Belgium and Portugal, who are members of treaty organizations, grant patents without doing any examinations or prior art searches. In the event that an infringement lawsuit is brought against you or the manufacturer, prior art will be determined by the court of law.

Some foreign countries may issue an opposition proceeding after the application is made public, which gives any opponents the opportunity to demonstrate prior art. Patent laws with regard to maintenance fees may be different depending on which treaty organization you file a foreign patent with. Other countries, such as Japan, require higher than normal fees to apply for a patent, and the registration process can take years. However, in countries where the application fees are higher and the registration process is longer, there will be less competition and greater market share due to the highly exclusive nature of that country's patent law. Many countries outside the U.S. consider any patent registration anywhere as proof of prior art. It is also wise to check patent laws for the sake of the type of product you are marketing, as some countries forbid foreign patent applications on drugs or computer programs.

Recent developments, however, plan to change the visage of U.S. patent laws. Under the Obama administration and the U.S. Senate, which approved the America Invents Act on March 9, 2011, the United States will be transitioning from a "first-to-invent" system to a "first-inventor-to-file" system. Under the first-to-invent system, patent law stated that patent claims would be defined by the documented date the invention was conceived and its reduction to practice. Under the first-inventor-to-file system, the right to a patent depends on the date the application was filed, regardless of the date of conception. Additional proposed changes to U.S. patent law through the America Invents Act include revisions to re-examination procedures and expanded methods of challenging a patent. Most countries with memberships to a treaty organization use the first-to-file system, and as American inventors expand their patents overseas, getting on the same system will help alleviate some of the confusion and problems involved in pursuing patents in other jurisdictions. Most stakeholders across the industrial spectrum are supporting this transition, including the National Association of Manufacturers, The Intellectual Property Owners Association, and the Coalition for 21st Century Patent Reform, which is a coalition of 50 major companies from 18 different industries. According to the Small Business & Entrepreneurship Council, "Patent harmonization among nations will make it easier, including less costly, for small firms and inventors to gain patent protection in other nations, which is critical to being able to compete internationally. By moving to a first-to-file system, small firms will in no way be disadvantaged, while opportunities in international markets will expand." For a full breakdown of the America Invents Act, go to **http://dpc.senate.gov/dpcdoc.cfm?doc_name=lb-112-1-9**.

Filing Methods

Five methods exist for filing an application in a foreign country. The first method is to file a U.S. patent application early, wait six months, file a non-Convention application before your product has been manufactured, file a PCT application a year after the original filing date, and file an EPO application 30 months after the original filing date. This is the most common method because it gives you all four types of applications within 30 months. The second method is to file a Chapter II prototype examination with PCT 19 months after a receiving a U.S. patent, and then file with the EPO 30 months after the original filing date. The third method is to file a U.S. patent application, and then file with the EPO and non-Convention counties one year after the original filing date. The fourth method is to file a U.S. patent and file with individual EPO countries (rather than the EPO) one year after the original filing date. The fifth method is to file a provisional patent and file a PCT to apply for a regular patent in the U.S. and a foreign patent.

If you file a provisional patent, you will receive a search report conducted by a patent agent. The PCT search report will include an analysis of your claims with respect to patentability and offer the chance to amend those claims if necessary. Three months from the date on the patent search report, it is feasible to elect for Chapter II. Electing for Chapter II will cause your claims to undergo a more rigorous examination and produce a more detailed search report. When 30 months have passed after the date of the provision patent application, you will be allowed to file national applications for individual countries in the EPO while filing for a RPA in the U.S. The reasons the first method (RPA and non-Convention filing) is the most common is because it is the least expensive way to obtain patents in the U.S. and abroad, and it gives you patent pending status 30 months from the original filing date. The steps involved in executing this most common method are:

1. Filing a RPA in the U.S.

2. Filing in non-Convention countries

3. Filing with the PCT one year after RPA

4. Waiting for the application to undergo search and examination

5. Waiting for the application to be sent to the EPO

6. Having the search report copies sent to countries you have applied to

7. Hiring a patent agent within 30 months of original date to send applications to countries that received search report copies

Foreign License Deals

Obtaining foreign license deals means being able to decide who will make and distribute your product overseas and ensures your product cannot be infringed upon. A foreign licensee should have a distribution network that can stock and ship while keeping delivery costs down. Although the cost of obtaining patent protection in the U.S. is anywhere from $2,000 to $10,000, the cost of obtaining foreign patents with the EPO, with ten-year renewal fees, can be anywhere from $27,000 to $35,000. The America Invents Act has been designed to bring those costs down, but the cash requirements will still be much higher than bringing a product to market in the U.S. Therefore, any foreign licensing deal with a manufacturer should cover filing costs in the agreement. A manufacturer is only willing to pay these costs if their marketing analysis matches your expectation of at least $500,000 in sales; otherwise, a grant of intellectual property rights is not worth pursuing.

When dealing with a manufacturer in a foreign license deal, ask the firm if they have ever licensed from a U.S. proprietor before. What you are looking for in this situation is experience because, as a licensor to international markets, you have none. If they have licensed from a U.S. proprietor before, ask for a full contact list of the licensors. Do not let them give you a select list, as some manufacturers may keep contacts from you that did not result in a satisfactory business relationship. Once you have the list, contact all of them, and ask them for their feedback and whether they were paid royalties in U.S. dollars. Many U.S. manufacturers have moved their manufacturing facilities overseas, so it will be no different in the case of a foreign licensee. In both cases, you will not be able to physically inspect the facility, so ask if they are willing to permit audit provisions in their agreements. An audit provision will allow you to inspect their books and receive reports on a periodic basis.

Another important issue to know before committing to a foreign licensing deal is the issue of jurisdiction. If you have to take a foreign licensee to court, the agreement must have a provision in which the licensee consents to U.S. jurisdiction. Because patent laws differ from country to country, you will want any litigation settled under U.S. law. If you do not have jurisdiction, the manufacturer can have the case decided in their country, which means you will have to learn the laws of that country and hire a foreign patent agent who knows those laws.

Just as you may want to provide a RPA provision to have rights revert to you after a certain period, you may want to do the same in a foreign license deal. To achieve this, a fixed term option should be included in the licensing agreement whereby the agreement terminates after a certain number of years. Under a right-to-terminate provision, have your patent agent list reasons that would trigger the clause, such as failure to pay royalties, failure to bring product to market, failure to keep confidentiality or trade secrets, violation of another part of the agreement, or failure to maintain liability insurance.

If some type of violation in the foreign license deal occurs, and you want to avoid the cost and duration of time associated with litigation, seek an arbitrator. An arbitrator is an official judge who will have the power to make a ruling over the dispute and award attorney fees to the victor if a provision has been stated in the license agreement. Three of the most popular international arbitration courts are The London Court of International Arbitration, The International Court of Arbitration of the International Chamber of Commerce, and the Arbitration Institute of the Stockholm Chamber of Commerce.

Foreign Market Analysis

The first step in determining whether the cost of entering foreign markets will bring enough returns is to do a marketing analysis for each nation you are targeting. Determining your invention's worth in a foreign market will depend on your ability to answer questions regarding the market's sales potential, future market, competition, production and operational costs, existing budget capacity of licensees, and patent protection. The first part of your analysis should be the potential customers. Can they afford to buy your product at the retail price you need to set in order to be profitable? Determining sales potential should answer this question, as well as the current environment in that market, which dictates the demand. If there is high demand, and consumers can afford your product, is there a high level of competition? The answer to this question depends on how mature the industry appears where your product will be sold.

Cursory analysis of a country's economic health can begin with an inspection of its gross national product (GNP). GNP measures a country's financial output of goods and services, personal consumption, government expenditures, private investment, inventory growth, and trade balance. The more developed the country's industry or GNP, the higher the level of competition. Higher GNP means more competition but higher potential sales. Lower GNP means lower competition but lower potential sales. GNP should be used as a starting point in your market research; never rely on it as your only data source. Emerging markets are markets in which competition has not yet peaked but is increasing. Your analysis of the future market will focus on when and how the level of competition peaks. So where can you find all this data?

The largest collector of data on foreign markets is the U.S. Department of Commerce (**www.commerce.gov**). In some cases, information is available free, but some data points can only be accessed for a fee. Government data sources are great for compiling information, but their numbers may be outdated by a couple of years. Never assume that a country's financial situation cannot change over a period of years. Most recently, Ireland went from a country that was building a vast new infrastructure in 2004 to needing a bailout from other countries in 2010. To efficiently collect data that is both broad and current, contact the Department of Commerce and ask for a list of local libraries that function as a government depository and where you might find Government Printing Office publications. Other government agencies that provide foreign market information include:

- FedWorld Information Network (**www.fedworld.gov**)
- International Trade Administration (**www.ita.doc.gov**)
- U.S. Small Business Administration (**www.sba.gov**)
- The World Bank (**www.worldbank.org**)

To ensure your market data is current and broad, cross reference government databases with information private companies that collect and disseminate market analysis about foreign markets on an ongoing basis. These groups include industry publications and trade organizations available in directories at your local library. Kompass is a foreign trade directory and can be found online at **http://us.kompass.com**. Other nongovernment organization websites include:

- Export/Import Bank of the United States: **www.exim.gov/about/links.cfm**
- Federation of International Trade Associations: **www.fita.org**
- Library Journal: **www.getcustoms.com/2004GTC/index.html**
- Global Edge: **http://globaledge.msu.edu/resourcedesk**
- International Chamber of Commerce: **www.chamber-of-commerce.com**
- Piers: **www.piers.com**

For a subscription fee, the National Trade Data Bank offers the most extensive information on trade in foreign countries. If you are looking for specific information on countries in the European Union, EU Business (**www.eubusiness.com**) carries a bevy of statistics on each individual economy. The European Patent Office will have this information as well, but because it is a government-run organization and may not have updated information, it should be used for cross-reference purposes only.

CASE STUDY: BUSINESS IN CHINA

Barbara Cary, president and CEO
Carey & Co.
P.O. Box 2291, Orinda, CA 94563
info@careyco.co
www.careyco.co
Phone: (925) 254-0765
Fax: (925) 254-4810

I am an author, entrepreneur, and mother who turned a $6,000 personal investment into a $47-million empire. Over a 20-year period, I have invented hundreds of products, own several patents, and currently run four companies. In the beginning, I knew nothing about mass merchandising, and I did all the wrong things. What I have learned is that the key to success is to get the order first, then find a manufacturer who is willing to get paid with funds from the sale. My most successful invention, the Hairagami, was inspired by my own frustration with the time it required to style my hair each day, so I developed a prototype that allowed women to put their hair in a bun within seconds. Shortly after, I was in business with QVC; sales for Hairagami have surpassed $10 million.

With the exception of the first 100 pieces, all of the manufacturing for the Hairagami was done in China. I have found that the Chinese are incredibly serious about doing business. They can get you a good price, excellent turnaround time, and fine quality. But in the process of getting Hairagami to market, I did encounter one very bad manufacturer. The manufacturer lied and told me that a supply of the clips was being shipped in a day or two.

I later learned they had not even purchased the steel to make them at that point, so I had to convince my buyer to give me a 60-day extension. After she reluctantly agreed, I found a new manufacturer, and they helped me recover from the experience with the bad manufacturer. When talking to a manufacturer about the status of an order, I do not want to hear, "Everything is fine." I want to hear specific details about each phase of the operation.

Many of the retailers I do business with are global, and most use some form of a planogram to structure their buying and selling of merchandise. A planogram is a more sophisticated yearly calendar. As a rule of thumb, I use eight months in advance of sale to figure when I will schedule a sales meeting. The reason I use eight months as a benchmark instead of a year is that most retailers re-evaluate their planogram six months into their planogram. At that time, they pull merchandise that is not selling off the shelf and look for replacement products. No two retailers and departments are on the same schedule. For example, Walmart, Kmart, Target, and CVS do not all designate the first week in March as the "buy hair accessories" period. That is why I build relationships with buyers and keep detailed notes of our conversations. They usually will tell you when they buy for their departments. In addition to a planogram, buyers also acquire products they sell "on promotion." In other words, items marketed specifically for holidays and special occasions. Seasons of the year are typically categories for promotion. For example, May through June is sometimes known as "Grads and Dads." Those promotional periods often correspond to designated areas in the stores where the items will be displayed. There is greater flexibility in the selling time with seasonal items than there are with the more rigid planograms.

In either case, getting information from buyers about their programs is essential. In my 2006 book, *The Carey Formula*, there is a planogram [on CD] and a promotion fill-in-the-blank form that charts the buying schedules of the buyers with whom you are working. I have found this to be an excellent organizational tool for planograms.

If your product is not merchandised in the correct place or displayed consistently where your end users can find it, you will not do well in a retail atmosphere. You must know how the stores in a particular chain are laid out. Strive to get your buyers and merchandisers to display your product in a year-round planogram program, so your customers can find it with ease and go back for repeat purchases.

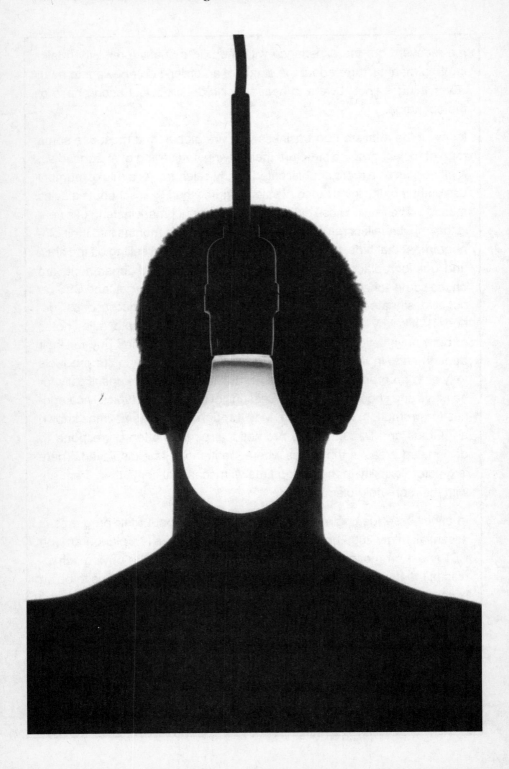

Conclusion

Our hope in writing this book is that you come away with a better understanding of what it takes to get your amazing invention on store shelves. If you started with the assumption that an innovative idea is all you need, you are certainly wiser now. A great inventor uses his or her ability to step outside the zone of comfort, learn new things, and solve complex and daunting tasks. As you attempt to bring an invention to market, you will learn all of the invention industry particulars, gain experience, and become the proprietor of the type of innovations that are helping to shape the future of the world. When new ways of doing things are needed, that is where you come in. You will be a master of innovation and product delivery, bringing new concepts into people's lives and enriching them in some meaningful way. Ideas, methods, and processes may become obsolete over time, but the demand for a new approach will always create new and exciting opportunities for professionals in the field. Should you choose to pursue it, the road ahead is both challenging and rewarding. We wish you luck.

Before closing this book, examine the questions at the end of this conclusion. They will help you determine just how ready you are to bring an invention to market. Whenever things seem tough, remember that creativity is the most important human resource of all. As American scholar Warren Bennis once said, "Any new idea, by definition, will not be accepted at first. It takes repeated attempts, endless demonstrations, and monotonous rehearsals before innovation can be accepted and internalized by an organization. This requires *courageous patience*."

Questions to ask after reading this book:

- Do I believe my product is marketable?
- Do I have the skills and knowledge to market it?
- Is it financially feasible to achieve all that it takes to create, distribute, and sell?
- Am I ready for such a large undertaking?
- Is there any information I am unsure about?
- Where can I find information (in this book or elsewhere) that will answer my remaining questions?

Appendices

Appendix A: List of Inventor Resources

- The National Inventor Fraud Center: **www.inventorfraud.com**

- Mom Inventors: **www.mominventors.com**

- Thomas Company's industry index: **www.thomasnet.com**

- USPTO: **www.uspto.gov**

- United Inventors' Association — nonprofit dedicated to educating inventors: **www.uiausa.org**

- Licensing Executives' Society — trade association of people interested in licensing intellectual property: **www.lesi.org**

- Inventor Spot — information, forums, and blog for inventors: **http://inventorspot.com/inventor_home**

- Inventor Ed — A nonprofit focused on invention marketing scams: **http://inventored.org**

- American Society of Inventors — inventors' association focused on education: **www.asoi.org**

- U.S. Copyright Office: **www.copyright.gov**

- InventNet — inventor information: **www.inventnet.com**

- Murray J. Atkins Library List of Inventor Web Resources: **http://library.uncc.edu/display/?dept=patents&format=open&page=1843**

- The Lemelson/MIT Foundation publishes the Inventor Handbook free online at **http://web.mit.edu/invent/h-main.html**.

Appendix B: Witness Form

Title of invention: _____

Description of invention _____

_____

_____

_____

_____

_____

_____

All accompanying illustrations, photographs, or visual art should be attached to this document. Rough sketch of invention may be drawn here:

Inventor's signature: _____   Date: _____

Witness has hereby read and understood the description above and attests to the accurateness of the statement.

Witness signature: _____ Date: _____

Address: _____

Phone: _____ E-mail: _____

Second witness (optional)

Witness has hereby read and understood the description above and attests to the accurateness of the statement.

Witness signature: _____ Date: _____

Address: _____

Phone: _____ E-mail: _____

Appendix C: Non-Disclosure Agreement Form

THIS AGREEMENT, made on [month/day/year] between _____
_____, ("Disclosing Party"), and _____
_____ ("Receiving Party").

BACKGROUND

The Disclosing Party and Receiving Party wish to discuss and exchange certain information related to the Disclosing party's "Invention," which both parties consider highly confidential and proprietary. Both parties are legally bound in consideration of the mutual agreements set forth herein, and hereby agree as follows:

DEFINITIONS

"**Invention**" shall refer to all information relating to conception, product, applications, components, technologies, and business topics.

"**Confidential Information**" shall refer to all information provided by Disclosing Party, whether it is in the form of written, oral, audio, visual, computerized, mechanical, prototypical material. Confidential Information shall also include all information divulged prior to the signing of this agreement.

Confidential Information shall not include:

    a) Public domain information at the time of the disclosure

b)   Documented information previously in the possession of Receiving Party at the time of disclosure

c)   Identical documented information previously disclosed to receiving party from a third party

USE OF CONFIDENTIAL INFORMATION

Receiving Party agrees to:

a)   Receive disclosed information and maintain confidentiality

b)   Not reproduce information without expressed written consent of disclosing party

c)   Not divulge or use information directly or indirectly to any person or business entity without expressed written consent of disclosing party

d)   Not use Confidential Information to design or create a similar apparatus, device, method, or system

e)   Make the best effort to protect Confidential Information

RETURN OF CONFIDENTIAL INFORMATION

All information provided by the Disclosing Party shall remain the property of the Disclosing Party. Receiving Party agrees to return all Confidential Information to Disclosing Party within 15 days of written demand by Disclosing Party. When the Receiving Party has finished reviewing the information provided by the Disclosing Party and has made a decision as to whether or not to work with the Disclosing Party, Receiving Party shall return all information to the Disclosing Party without retaining any copies.

NON-ASSIGNABLE

This agreement shall be non-assignable by the Receiving Party unless prior written consent of the Disclosing Party is received. If this Agreement is assigned or otherwise transferred, it shall be binding on all successors and assigns.

No License

Neither party grants, either expressly or by implication, any right or license to any patent, trade secret, invention, trademark, copyright, or other intellectual property right.

Provisions Separable

The provisions of this Agreement are independent of and separable from each other, and no provision shall be affected or rendered invalid or unenforceable by virtue of the fact that for any reason any other or others of them may be invalid or unenforceable in whole or in part.

ENTIRE AGREEMENT

This Agreement sets forth all of the covenants, promises, agreements, conditions and understandings between the parties and there are no covenants, promises, agreements or conditions, either oral or written, between them other than herein set forth. No subsequent alteration, amendment, change, or addition to this Agreement shall be binding upon either party unless reduced in writing and signed by them.

Arbitration

Any controversy, breech, or claim arising out of or relating to this Agreement shall be resolved by arbitration conducted by the Commercial Division of the American Arbitration Association. Any arbitration award shall be final and binding, and judgment upon the award rendered pursuant to such arbitration may be entered in any court of proper jurisdiction. Either party may seek and obtain temporary injunctive relief from any court of competent jurisdiction against any improper disclosure of the Confidential Information.

IN WITNESS OF THEIR AGREEMENT, both parties have agreed to conditions effective the month, day, and year first written above.

Signature (Disclosing Party): _____

Signature (Receiving Party): _____

Appendix D: Proprietary Submission Statement

[COMPANY] welcomes your new product ideas and inventions as well as your interest in our company. Because our in-house Research and Development team is constantly working on new products and improvements of existing products, certain precautions must be made in accepting new ideas and inventions for evaluation. In that regard, please realize that an idea that is new to you may have already been submitted to us by another party or conceived in-house by our own engineers and researchers. Therefore, to avoid possible confusion or misunderstandings as to the origin of an idea, a definitive understanding must be reached before [Company] can agree to review ideas from persons who are not employees of [Company]. Accordingly, [Company] will evaluate your submission only upon your review and acknowledgment of the following:

1. I hereby represent that I am the sole originator of the idea/ invention, that I own the idea/invention, and that I have the legal right to negotiate with [Company] concerning the submitted idea/ invention.

2. I realize that my idea/invention may be partially or wholly within the public domain, that employees of [Company] may have worked, or may be working on the same idea or invention, or that [Company] may have received similar information from others.

3. Absent a written agreement with [Company] to the contrary, I agree that I will rely solely on my rights under the Patent, Trademark and Copyright laws of the United States and that consideration of my submission to [Company] shall in no way impair [Company]'s right to contest the validity of my submission.

4. Absent a written agreement with [Company] to the contrary, I agree that [Company] has the right to disclose the idea/invention

submission to various employees and to those outside of its employ to determine the value of the submission. Therefore, the disclosure is not confidential and no confidential relationship is entered into because [Company] is considering the submission.

5. Absent a written agreement with [Company] to the contrary, I understand that [Company] is under no obligation to return submitted materials and if returned, accepts no responsibility for the safe arrival, handling, or return of the submitted materials.

6. I understand that [Company] assumes no obligation to evaluate the submission or to do more than to indicate whether or not the company is interested in pursuing the submission.

7. No agreement for compensation is to be implied from [Company]'s consideration or review of my idea/submission. After evaluation if [Company] decides to purchase, license, patent, or otherwise enter into an agreement with me, the extent of compensation will be determined by a mutual written agreement.

ACCEPTED AND AGREED BY:

Signature: _____ Date: _____

Printed name: _____

Address: _____ City/State/Zip: _____

Phone: _____ Fax: _____

E-mail: _____

Appendix E: Business Proposal Form

Submission Prepared for: _____

Submission Prepared by: _____

Description:

[Inventor name] is interested in [services offered by industry], and we would like to give [name of prospective licensee] the opportunity to purchase rights to manufacture or distribute our latest invention, which we call [name of invention]. This new invention [description of its function, services, and benefit to customers].

We are pleased to submit, for the approval of [Company], herein after referred to as Client, the following proposal for the production of [Invention], herein after referred to as the Product, in [Location of manufacturer], herein after referred to as the Territory. The Product is defined as [description of product].

Attachments contain detailed Product specifications and production information.

The project will commence upon acceptance of this agreement. Client shall complete solicitation and compilation of listings within approximately [timeframe].

Approximate market share obtained should equal at least [percentage] of the total market share identified in the Territory. Distribution should be completed no more than [timeframe] after manufacturing has been completed. Finished Product should be completed within [timeframe] of all components delivered to the production facility. As part of this license, [inventor] will furnish to Client:

- Permission to use the trademark, as well as the registered trademark graphic logo, in all promotional materials on behalf of the project, and for a period of one year following the date of delivery of the finished Product
- Written operations, including step-by-step instructions for how to develop Product
- All layout and design elements in conjunction with the established Product specifications
- Training and consultation on development of the Product for a fee of [x amount] per hour

In return, Client will provide the following:

- License Fee in the amount of [x amount] due upon acceptance of this agreement
- Production Services Fee in the amount of [x] due as condition of delivery of Product
- Diligent effort to obtain required market share
- Diligent effort to conform to training instructions
- Timely submission of draft versions of Product for review
- Receipt and removal of finished Product from production facility
- Distribution of Product throughout the Territory. Suggested distribution points include [list].
- [Inventor] also requests the right to solicit and place paid advertisements as indicated in the attached Product specification.

Patents:

[Inventor] has applied for the following patents and expects to receive these patents within [timeframe]. Complete patent descriptions will be disclosed to those parties seriously interested in licensing the rights to manufacture or distribute these products.

[List patent types and description of claims]

Thank you for the opportunity to submit this proposal. We believe that your organization and the community will derive great benefit and value from this product. We will do everything in our power to ensure your satisfaction with this project.

ACCEPTED BY: _____

DATE: _____

CLIENT: _____

TITLE: _____

AUTHORIZED SIGNATURE: _____

Appendix F: License Agreement Form

I. Introduction

This License Agreement is between [Inventor] (herein referred to as "Licensor"), and [Company] (referred to as "Licensee"). Licensee desires to license certain rights in the invention. Both parties agree to transfer rights as follows: [*check the following appropriate options*]

[ ] **Patent issued**

U.S. Patent No(s):

[ ] **Patent Not Yet Issued**

U.S. Patent Application No(s):

[ ] **Patents and Improvements**

The Property is defined as the invention(s) described in U.S. Patent No(s): [insert patent number(s)]; any changes or improvements that shall be based on the patent(s); any patent applications corresponding to the above-described patent(s); and patent applications that are issued, filed, or to be filed in any and all foreign countries.

[ ] **Copyright, Trade Secrets, and Trademarks: No Patents**

The Property refers to all proprietary rights, including copyrights, trade secrets, formulas, research data, know-how, and specifications related to the [Invention] as well as any trademark rights and associated goodwill.

[ ] **Copyright or Trademark Registration**

The Property refers to copyrights as embodied in

Copyright Registration No:

Trademark Registration No:

[ ] Patents and Copyright, Trade Secrets, and Trademarks

Refers to all inventions described in [number of patents] and to all other proprietary rights, including copyrights, trade secrets, formulas, research data, know-how, and specifications related to the invention, as well as any trademark rights and associated goodwill.

II. Licensed Products

Both Parties agree upon:

Licensed Products are defined as the Licensee products incorporating the Property. Licensed Products are defined as any products sold by the Licensee that incorporate the Property. A Licensed Process is any commercial application of the Process by the Licensee.

III. Grant of Rights

Licensor grants to Licensee [exclusive or non-exclusive] license to make, use, and sell the Property solely in association with the manufacture, sale, use, promotion, or distribution of the Licensed Products.

IV. Sublicense

Both Parties agree upon [*choose one of the options below*]

[ ] Sublicense

Licensee may sublicense the rights granted in accordance with this agreement provided Licensor's prior written consent, and review of revenue or royalty payment from Sublicensing Revenues. Any sublicense granted in violation of this provision shall be void.

[ ] Consent to Sublicense Not Unreasonably Withheld

Licensee may sublicense the rights granted pursuant to this agreement provided Licensee obtains Licensor's prior written consent to such sublicense (Licensor's consent to any sublicense shall not be unreasonably withheld), and Licensor receives such revenue or royalty payment as provided in the Payment section below. Any sublicense granted in violation of this provision shall be void.

V. Reservation of Rights

Both Parties agree upon [*choose appropriate options below*]:

[ ] All Rights Reserved

Licensor expressly reserves all rights other than those being conveyed or granted in this Agreement.

[ ] Reservation of Rights Expressly Excluding a Particular Industry

Licensor expressly reserves all rights other than those being conveyed or granted in this agreement, including the right to license the Properties in [territory].

VI. Territory

Both Parties agree upon [*choose appropriate options below*]:

[ ] Statement of Territory

The rights granted to Licensee are limited to [Territory]

[ ] Limiting Cross-Territory Sales

The rights granted to Licensee are limited to [Territory]. Licensee shall not make, use, or sell the Licensed Products or any products that are similar to the Licensed Products in any country outside the Territory and will not knowingly sell the Licensed Products to persons who intend to resell them in a country outside the Territory.

VII. Term

Both Parties agree upon [*choose appropriate options below*]:

[ ] Specified Term and Renewal Rights

This Agreement shall be effective upon [date of signature] and shall extend for a period of [initial term agreed to]. Following the Initial Term, this agreement may be renewed by Licensee under the same terms and conditions for [renewal terms] for [renewal period] provided Licensees expressed written intention to renew this agreement within 30 days before the expiration of the current term.

[ ] Term for the Length of Patent Only

This Agreement shall effective upon [date of signature] and shall expire with the expiration of the last patent or patent application unless terminated pursuant to the provisions of this Agreement.

[ ] Short Term With Renewal Rights Based Upon Sales

This Agreement is effective upon [date of signature] and shall extend for a period of [timeframe] and thereafter, may be renewed by Licensee under the same terms and conditions for [timeframe] provided that:

(a) Licensee provides written notice of its intention to renew this agreement within 30 days before the expiration of the current term

(b) Licensee has met the sales requirements

(c) In no event shall the Agreement extend longer than the date of expiration of the last expiring patent(s) in the definition of the Property.

[ ] No Patents; Indefinite Term

This Agreement is effective upon [date of signature] and shall continue until terminated pursuant to a provision of this Agreement.

[ ] Fixed Yearly Term

This Agreement is effective upon [date of signature] and shall continue for [timeframe] unless sooner terminated pursuant to a provision of this Agreement.

[ ] Term for as Long as Licensee Sells Licensed Products

This Agreement is effective upon [date of signature] for Licensed Product is offered in commercial quantities unless terminated pursuant to a provision of this Agreement.

VIII. Royalties

All royalties shall accrue when items are sold, shipped, distributed, billed, or paid for. Royalties shall also be paid by the Licensee to Licensor on all items, even if not billed to individuals or companies which are affiliated with, associated with, or subsidiaries of Licensee.

IX. Net Sales

"Net Sales" are defined as Licensee's gross sales less quantity discounts and returns actually credited. A quantity discount is a discount made at the time of shipment. No deductions shall be made for cash or other discounts, for commissions, for uncollectible accounts, or for fees or expenses of any kind that may be incurred by the Licensee in connection with the Royalty payments.

Both Parties agree to: [*choose appropriate options*]

[ ] **Advance Against Royalties**

Licensee agrees to pay to Licensor upon execution of this Agreement the sum of [x amount].

X. Licensed Product Royalty

Both Parties agree to [*choose appropriate options below*]:

[ ] **Royalty on Net Sales**

Licensee agrees to pay a Royalty of [percentage of all Net Sales] of the Licensed Products.

[ ] **Per-Unit Royalties**

Licensee agrees to pay a Royalty of [x amount] per Unit.

[ ] **Hybrid Royalty; Patent and Non-patented Rights**

Licensee agrees to pay a Royalty of [percentage of all Net Sales] of the Licensed Products. In the event that a patent does not issue or an issued patent expires or is terminated, the percentage for such patent or pending patent shall be subtracted from the Licensed Product Royalty. The Licensed Product Royalty shall be adjusted accordingly.

% of the Royalty for the license of the Patent No. _____

% of the Royalty for the license of Pending Patent No. _____

% of the Royalty for the license of trade secrets [*or* trademarks]

[ ] **Guaranteed Minimum Annual Royalty Payment**

In addition to any other advances or fees, Licensee shall pay an annual guaranteed Royalty as follows: [terms of GMAR]. The GMAR shall be paid to Licensor annually on [date]. The GMAR is an advance against Royalties for the 12-month period commencing upon payment. Royalty payments based on Net Sales made during any year of this Agreement shall be credited against the GMAR due for the year in which such Net Sales were made. In the event that annual Royalties exceed the GMAR, Licensee shall pay the difference to Licensor. Any annual Royalty payments in excess of the GMAR shall not be carried forward from previous years or applied against the GMAR.

[ ] License Fee

As a nonrefundable fee for executing this license, Licensee agrees to pay to Licensor upon execution of this Agreement the sum of $_____.

[ ] Royalties on Spin-Offs

Licensee agrees to pay a Royalty of [percentage for all Net Sales] of "Spin-Off Products." A Spin-Off Product is any product that is derived from, based on, or adapted from the Licensed Product, provided that if the product uses the Property, it shall be considered a Licensed Product and not a Spin-Off Product.

[ ] Adjustment of Royalties for Third-Party Licenses

In the event that any Licensed Product (or other items for which Licensee pays Royalties to Licensor) incorporates third-party character licenses, endorsements, or other proprietary licenses, Licensor agrees to adjust the Royalty rate to [royalty percentage] for such third-party licenses. Licensee shall notify Licensor of any such third-party licenses prior to manufacture. Third-party licenses shall not include licenses accruing to an affiliate, associate, or subsidiary of Licensee.

[ ] F.O.B. Royalties

Licensee agrees to pay the Royalty of [percentage] for all F.O.B. sales of Licensed Products.

[ ]   Sublicensing Revenues

In the event of any sublicense of the rights granted pursuant to this Agreement, Licensee shall pay to Licensor [percentage] of all sublicensing revenues.

Payments and Statements to Licensor

Within 30 days after the end of each calendar quarter, an accurate statement of Net Sales of Licensed Products along with any Royalty payments or sublicensing revenues due to Licensor shall be provided to Licensor.

XI. Audit

Licensee shall keep accurate books of account and records covering all transactions relating to the license granted in this Agreement, and Licensor or its duly authorized representatives shall have the right upon five days' prior written notice, and during normal business hours, to inspect and audit Licensee's records relating to the Property licensed under this Agreement. Licensor shall bear the cost of such inspection and audit, unless the results indicate an underpayment greater than [x amount] for any six-month period. In that case, Licensee shall promptly reimburse Licensor for all costs of the audit along with the amount due with interest on such sums. Interest shall accrue from the date the payment was originally due, and the interest rate shall be 1.5% per month, or the maximum rate permitted by law, whichever is less. All books of account and records shall be made available in the United States and kept available for at least two years after the termination of this Agreement.

XII. Late Payment

Time is of the essence with respect to all payments to be made by Licensee under this Agreement. If Licensee is late in any payment provided for in this Agreement, Licensee shall pay interest on the payment from the date due until paid at a rate of 1.5% per month, or the maximum rate permitted by law.

XIII. Licensor Warranties

Licensor warrants that it has the power and authority to enter into this Agreement and has no knowledge as to any third-party claims regarding the proprietary rights in the Property that would interfere with the rights granted under this Agreement.

XIV. Indemnification by Licensor

Both Parties agree upon [*choose appropriate options below*]:

[ ]   Statement of Licensor Indemnification

Licensor shall indemnify Licensee and hold Licensee harmless from any damages and liabilities arising from any breach of Licensor's warranties as defined in Licensor Warranties, above, provided (a) such claim, if sustained, would prevent Licensee from marketing the Licensed Products or the Property; (b) such claim arises solely out of the Property as disclosed to the Licensee, and not out of any change in the Property made by Licensee or a vendor, or by reason of an off-the-shelf component or by reason of any claim for trademark infringement; (c) Licensee gives Licensor prompt written notice of any such claim; (d) such indemnity shall only be applicable in the event of a final decision by a court of competent jurisdiction from which no right to appeal exists; and (e) the maximum amount due from Licensor to Licensee under this paragraph shall not exceed the amounts due to Licensor under the Payment section from the date that Licensor notifies Licensee of the existence of such a claim.

[ ] Licensor Indemnification

Licensor shall indemnify Licensee and hold Licensee harmless from any damages and liabilities (including reasonable attorneys' fees and costs) arising from any breach of Licensor's warranties as defined in Licensor Warranties, above, provided (a) such claim, if sustained, would prevent Licensee from marketing the Licensed Products or the Property; (b) such claim arises solely out of the Property as disclosed to the Licensee, and not out of any change in the Property made by Licensee or a vendor, or by reason of an off-the-shelf component or by reason of any claim for trademark infringement; (c) Licensee gives Licensor prompt written notice of any such claim; (d) such indemnity shall only be applicable in the event of a final decision by a court of competent jurisdiction from which no right to appeal exists; and (e) the maximum amount due from Licensor to Licensee under this paragraph shall not exceed the amounts due to Licensor under the Payment section from the date that Licensee notifies Licensor of the existence of such a claim. The maximum amount due from Licensor to Licensee under this paragraph shall not exceed _____ *[50 to 100]* _____ % of the amounts due to Licensor under the Payment section _____ *[if you have numbered the sections of your agreement, include the number of the payment section]* _____ from the date that Licensor notifies Licensee of the existence of such a claim. After the commencement of a lawsuit against Licensee that comes within the scope of this paragraph, Licensee may place _____ *[same percentage as listed above]* _____ % of the royalties thereafter due to Licensor under the Payment section in a separate interest-bearing fund that shall be called the "Legal Fund." Licensee may draw against such Legal Fund to satisfy all of the reasonable expenses of defending the suit and of any judgment or settlement made in regard to this suit. In the event the Legal Fund is insufficient to pay the then-current defense obligations, Licensee may advance monies on behalf of the

Legal Fund and shall be reimbursed as payments are credited to the Legal Fund. Licensor's liability to Licensee shall not extend beyond the loss of its royalty deposit in the Legal Fund. After the suit has been concluded any balance remaining in the Legal Fund shall be paid to Licensor, and all future royalties due to Licensor shall be paid to Licensor as they would otherwise become due. Licensee shall not permit the time for appeal from an adverse decision on a claim to expire.

XV. Warranties

Licensee warrants that it will use its best commercial efforts to market the Licensed Products and that their sale and marketing shall be in conformance with all applicable laws and regulations, including but not limited to all intellectual property laws.

XVI. Indemnification

Licensee shall indemnify Licensor and hold Licensor harmless from any damages and liabilities (including reasonable attorneys' fees and costs) (a) arising from any breach of Licensee's warranties and representation as defined in the Licensee Warranties, above; (b) arising out of any alleged defects or failures to perform of the Licensed Products or any product liability claims or use of the Licensed Products; and (c) any claims arising out of advertising, distribution, or marketing of the Licensed Products.

XVII. IP Protection

Licensor may, but is not obligated to, seek in its own name and at its own expense, appropriate patent, trademark, or copyright protection for the Property. Licensor makes no warranty with respect to the validity of any patent, trademark, or copyright that may be granted. Licensor grants to

Licensee the right to apply for patents on the Property or Licensed Products provided that such patents shall be applied for in the name of Licensor and licensed to Licensee during the Term and according to the conditions of this Agreement. Licensee shall have the right to deduct its reasonable out-of-pocket expenses for the preparation, filing, and prosecution of any such U.S. patent application (but in no event more than $5,000) from future royalties due to Licensor under this Agreement. Licensee shall obtain Licensor's prior written consent before incurring expenses for any foreign patent application.

XVIII. IP Laws

The license granted in this Agreement is conditioned on Licensee's compliance with the provisions of the intellectual property laws of the United States and any foreign country in the Territory. All copies of the Licensed Product as well as all promotional material shall bear appropriate proprietary notices.

XIX. Infringement Lawsuits Against Third Parties

In the event that either party learns of imitations or infringements of the Property or Licensed Products, that party shall notify the other in writing of the infringements or imitations. Licensor shall have the right to commence lawsuits against third persons arising from infringement of the Property or Licensed Products. In the event that Licensor does not commence a lawsuit against an alleged infringer within 60 days of notification by Licensee, Licensee may commence a lawsuit against the third party. Before filing suit, Licensee shall obtain the written consent of Licensor to do so, and such consent shall not be unreasonably withheld. Licensor will cooperate fully and in good faith with Licensee for the purpose of securing and preserving Licensee's rights to the Property. Any recovery (including, but not limited to, a judgment, settlement, or licensing agreement included as resolution of an infringement dispute) shall be divided equally between the parties after deduction and payment of reasonable attorneys' fees to the party bringing the lawsuit.

XX. Exploitation

Licensee agrees to manufacture, distribute, and sell the Licensed Products in commercially reasonable quantities during the term of this Agreement and to commence such manufacture, distribution, and sale within [timeframe].

XXI. Samples and Quality Control

Licensee shall submit a reasonable number of production samples of the Licensed Product to Licensor to ensure that the Licensed Product meets Licensor's quality standards. In the event that Licensor fails to object in writing within ten business days after the date of receipt, the Licensed Product shall be deemed to be acceptable. At least once during each calendar year, Licensee shall submit two production samples of each Licensed Product for review. The quality standards applied by Licensor shall be no more rigorous than the quality standards applied by Licensee to similar products.

XXII. Insurance

Licensee shall, throughout the Term, obtain and maintain, at its own expense, standard product liability insurance coverage, naming Licensor as additional named insured. Such policy shall (a) be maintained with a carrier having a Moody's rating of at least B; and (b) provide protection against any claims, demands, and causes of action arising out of any alleged defects or failure to perform of the Licensed Products or any use of the Licensed Products. The amount of coverage shall be a minimum of [x dollars] with no deductible amount for each single occurrence for bodily injury or property damage. The policy shall provide for notice to the Agent and Licensor from the insurer by registered or certified mail in the event of any modification or termination of insurance. Licensee shall furnish Licensor and Agent a certificate from its product liability insurance carrier evidencing insurance coverage in favor of Licensor and in no event shall Licensee distribute the Licensed Products before the receipt by the Licensor of evidence of insurance. The provisions of this section shall survive termination for three years.

XXIII. Confidentiality

The parties acknowledge that each may be furnished or have access to confidential information that relates to each other's business (the "Confidential Information"). In the event that Information is in written form, the disclosing party shall label or stamp the materials with the word "Confidential" or some similar warning. In the event that Confidential Information is transmitted orally, the disclosing party shall promptly provide a writing indicating that such oral communication constituted Information. The parties agree to maintain the Confidential Information in strictest confidence for the sole and exclusive benefit of the other party and to restrict access to such Confidential Information to persons bound by this Agreement, only on a need-to-know basis. Neither party, without prior written approval of the other, shall use or otherwise disclose to others, or permit the use by others, of the Confidential Information.

XXIV. Termination

Both parties agree upon [*choose one of the options below*]:

[ ]   Initial Term With Renewals From the Term Section

This Agreement terminates at the end of initial term unless renewed by Licensee under the terms and conditions as provided in the Term section of this Agreement.

[ ]   Fixed Term

This Agreement shall terminate at the end of [x] years unless terminated sooner under a provision of this Agreement.

[ ]   Term Based Upon Length of Patent Protection

This Agreement shall terminate with the expiration of the longest-living patent (or patents) or last-remaining patent application, unless terminated sooner under a provision of this Agreement.

[ ]   Initial Term With Renewals

This Agreement terminates at the end of the initial term unless renewed by Licensee under the same terms and conditions for consecutive two-year periods, provided that Licensee provides written notice of its intention to renew this agreement within 30 days prior to expiration of the current term. In no event, shall the Agreement extend longer than the date of expiration of the longest-living patent(s) or last-remaining patent application as listed in the definition of the Property.

[ ] Termination at Will: Licensee's Option

Upon 90 days' notice, licensee may, at its sole discretion, terminate this agreement by providing notice to the licensor.

XXV. Licensor's Right to Terminate

Licensor shall have the right to terminate this Agreement for the following reasons:

(a) Licensee fails to pay Royalties when due or fails to accurately report Net Sales, as defined in the Payment section of this Agreement, and such failure is not cured within 30 days after written notice from the Licensor;

(b) Licensee fails to introduce the product to market by _____ [*insert date by which Licensee must begin selling Licensed Products*] _____ or to offer the Licensed Products in commercially reasonable quantities during any subsequent year;

(c) Licensee fails to maintain confidentiality regarding Licensor's trade secrets and other Information;

(d) Licensee assigns or sublicenses in violation of the Agreement; or

(e) Licensee fails to maintain or obtain product liability insurance as required by the provisions of this Agreement.

[ ] Terminate as to Territory Not Exploited

Licensor shall have the right to terminate the grant of license under this Agreement with respect to any country or region included in the Territory in which Licensee fails to offer the Licensed Products for sale or distribution or to secure a sublicensing agreement for the marketing, distribution, and sale of the product within two years of the Effective Date.

XXVI. Effect of Termination

Upon termination of this Agreement, all Royalty obligations as established in the Payment section shall immediately become due. After the termination of this license, all rights granted to Licensee under this Agreement shall terminate and revert to Licensor, and Licensee will refrain from further manufacturing, copying, marketing, distribution, or use of any Licensed Product or other product that incorporates the Property. Within 30 days after termination, Licensee shall deliver to Licensor a statement indicating the number and description of the Licensed Products that it had on hand or was in the process of manufacturing as of the termination date. Licensee may dispose of the Licensed Products covered by this Agreement for a period of three months after termination or expiration, except that Licensee shall have no such right in the event this agreement is terminated according to the Licensor's Right to Terminate, above. At the end of the post-termination sale period, Licensee shall furnish a royalty payment and statement as required under the Payment section. Upon termination, Licensee shall deliver to Licensor all tooling and molds used in the manufacture of the Licensed Products. Licensor shall bear the costs of shipping for the tooling and molds.

XXVII. Survival

The obligations of Sections _____ *[insert section names or numbers that will survive termination]* _____ shall survive any termination of this Agreement.

XXVIII. Attorneys' Fees and Expenses

The prevailing party shall have the right to collect from the other party its reasonable costs and necessary disbursements and attorneys' fees incurred in enforcing this Agreement.

XXIX. Dispute Resolution

Both Parties agree upon [choose the appropriate options below]:

[ ] Arbitration

If a dispute arises under or relating to this Agreement, the parties agree to submit such dispute to binding arbitration in [location] or another location mutually agreeable to the parties. The arbitration shall be conducted on a confidential basis pursuant to the Commercial Arbitration Rules of the American Arbitration Association. Any decision or award as a result of any such arbitration proceeding shall be in writing, shall provide an explanation for all conclusions of law and fact, and shall include the assessment of costs, expenses, and reasonable attorneys' fees. An arbitrator experienced in invention licensing law shall conduct any such arbitration and shall include a written record of the arbitration hearing. The parties reserve the right to object to any individual who shall be employed by or affiliated with a competing organization or entity. An award of arbitration may be confirmed in a court of competent jurisdiction.

[ ] Mediation and Arbitration

The parties agree that every dispute or difference between them, arising under this Agreement, shall be settled first by a meeting of the parties attempting to confer and resolve the dispute in a good faith manner. If the parties cannot resolve their dispute after conferring, any party may require the other parties to submit the matter to nonbinding mediation, using the services of an impartial professional mediator approved by all parties. If the parties cannot come to an agreement following mediation, the parties agree to submit the matter to binding arbitration at a location mutually agreeable to the parties. The arbitration shall be conducted on a confidential basis pursuant to the Commercial Arbitration Rules of the American Arbitration Association. Any decision or award as a result of any such arbitration proceeding shall include the assessment of costs, expenses, and reasonable attorneys' fees and shall include a written record of the proceedings and a written determination of the arbitrators. Absent an agreement to the contrary, any such arbitration shall be conducted by an arbitrator experienced in intellectual property law. The parties reserve the right to object to any individual who shall be employed by or affiliated with a competing organization or entity. In the event of any such dispute or difference, either party may give to the other notice requiring that the matter be settled by arbitration. An award

262 How to Get Your Amazing Invention on Store Shelves

of arbitration shall be final and binding on the parties and may be confirmed in a court of competent jurisdiction.

[ ] Alternative Dispute Resolution

If a dispute arises and cannot be resolved by the parties, either party may make a written demand for formal resolution of the dispute. The written request will specify the scope of the dispute. Within 30 days after such written notice, the parties agree to meet, for one day, with an impartial mediator and consider dispute resolution alternatives other than litigation. If an alternative method of dispute resolution is not agreed upon within 30 days of the one-day mediation, either side may start litigation proceedings.

XXX. Governing Law

This Agreement shall be governed in accordance with the laws of [state].

XXXI. Jurisdiction

Each party (a) consents to the exclusive jurisdiction and venue of the federal and state courts located in [state] in any action arising out of or relating to this agreement; (b) waives any objection it might have to jurisdiction or venue of such forums or that the forum is inconvenient; and (c) agrees not to bring any such action in any other jurisdiction or venue to which either party might be entitled by domicile or otherwise.

XXXII. Waiver

The failure to exercise any right provided in this Agreement shall not be a waiver of prior or subsequent rights.

XXXIII. Invalidity

If any provision of this Agreement is invalid under applicable statute or rule of law, it is to be considered omitted, and the remaining provisions of this Agreement shall in no way be affected.

XXXIV. Entire Understanding

This Agreement expresses the complete understanding of the parties and supersedes all prior representations, agreements, and understandings, whether written or oral. This Agreement may not be altered except by a written document signed by both parties.

XXXV. Attachments and Exhibits

The parties agree and acknowledge that all attachments, exhibits, and schedules referred to in this Agreement are incorporated in this Agreement by reference.

XXXVI. Notices

Any notice or communication required or permitted to be given under this Agreement shall be sufficiently given when received by certified mail, sent by facsimile transmission, or overnight courier.

XXXVII. No Joint Venture

Nothing contained in this Agreement shall be construed to place the parties in the relationship of agent, employee, franchisee, officer, partners, or joint venturers. Neither party may create or assume any obligation on behalf of the other.

XXXV. Assignability

Both Parties agree upon [*choose appropriate options below*]:

[ ] **Statement of Assignability**

Licensee may not assign or transfer its rights or obligations pursuant to this Agreement without the prior written consent of Licensor. Any assignment or transfer in violation of this section shall be void.

[ ] **Consent Not Unreasonably Withheld**

Licensee may not assign or transfer its rights or obligations pursuant to this Agreement without the prior written consent of Licensor. Such consent shall not be unreasonably withheld. Any assignment or transfer in violation of this section shall be void.

[ ] **Consent Not Needed for Licensee Affiliates or New Owners**

Licensee may not assign or transfer its rights or obligations pursuant to this Agreement without the prior written consent of Licensor. However, no consent is required for an assignment or transfer that occurs (a) to an entity in which Licensee owns more than 50% of the assets, or (b) as part of a transfer of all or substantially all of the assets of Licensee to any party. Any assignment or transfer in violation of this section shall be void. Each party has signed this Agreement through its authorized representative. The parties, having read this Agreement, indicate their consent to the terms and conditions by their signatures below.

LICENSOR:

_____

Signature

_____

Name/Title

_____

Company

_____

Date

LICENSEE:

_____

Signature

_____

Name/Title

_____

Company

_____

Date

Appendix G: Sample Sell Sheet

Discount Schedule for Product X

Brief description of Product:

Product Reviews from Trade Journals:

Discounts:
3–299 units: -40%
300–499 units: -50%
500–up: -55%

Credit: Orders more than $50 must be prepaid or send bank and three trade references.

Shipping: Product is best shipped via UPS or USPS Priority Mail. We can ship via truck or Federal Express but do not recommend them for long distances because the rates are considerably higher. Shipping is FOB Ocala, Florida. Shortages or nonreceipt must be reported to us within 30 days of the ship date.

Resale Numbers: Florida dealers must mention their resale number with their order. Orders may be sent to the address below. Telephone orders may be made to xxx-xxx-xxxx. 9:00 a.m. to 5:00 p.m., Monday through Friday, Eastern Time (or to the answering machine after hours).

Secure online purchases can be ordered through *(website)* at any time.

Return Policy

PRODUCT X IS RETURNABLE. If Product X is not moving in your market, we want to get it back before a new version makes it obsolete. Thank you for giving it a chance on your valuable shelf space. Our return period is normally between 90 DAYS AND ONE YEAR of the manufacturer's invoice date.

RETURN PERMISSION MUST BE REQUESTED so that we can issue detailed packing and shipping instructions. This is your authorization and the instructions are below.

NOTICE OF SHORTAGE OR NONRECEIPT must be made within 30 days of shipping/invoice date for domestic shipments, 60 days for foreign.

PRODUCTS DAMAGED IN TRANSIT are not the responsibility of the manufacturer. Please make claim to the carrier. Returns must be accompanied by your packing slip listing

QUANTITY, ORIGINAL INVOICE NUMBER and INVOICE DATE. Products returned with this information will be credited with 100% of the invoice price minus shipping. Otherwise, it will be assumed that the original discount was 55%. All products should be returned to their original source (for example, the manufacturer or distributor).

ROUTING: Ship products via parcel post prepaid or UPS prepaid to [address].

To qualify for a refund, returned products must arrive in good Resalable condition. If they are not now resalable, please do not return them. If you are not willing to package them properly for the return trip, please do not waste your time and postage.

To package the products so that they will survive the trip, we suggest you wrap them the same way they were sent to you.

Appendix H: Consumer Questionnaire

Have you ever purchased any _____ products? (List product category)

What types of _____ products have you purchased in the last 12 months?

Why did you purchase these products?

Where did you purchase these products?

About how much did you pay for these products?

Would you use a product that _____?
(Describe benefits of your product)

Would you buy such a product?

How much would you expect to pay for a product like that?

What would be the most important features, functions, purposes, or benefits of such a product?

Appendix I: Distribution Chain Interview Questions

Some of these questions are geared toward retail store clerks; others are intended for store purchasing staff, distributor sales staff, distributor purchasing staff, manufacturer sales staff, or several of the above. Modify them as needed for each layer of the distribution chain. Work your way backward up the distribution chain from retail to manufacturing, from sales to purchasing, asking these type of questions at each level.

Begin all of these interviews by introducing yourself:

Hi, I am ___(your name)___. I am a product developer working on a (<u>one phrase description of product</u>, for example "a bath toy for infants"). I am conducting research on product distribution. Do you have a minute or two?

- How many products of this type does your store carry?
- Are they all displayed in one place, or do they fit into more than one display area in the store?
- What is the range of prices for these products in your store?
- How does price affect consumer purchasing of this type of product?
- What type of consumers purchase most of these products?
- What features are consumers looking for in this type of product?
- How do packaging and advertising affect sales of this type of product?
- How many units of any one of these products does your store sell in a month or year? How often do you reorder or restock these items?
- What sort of trends have you noticed in the industry?
- What sets one product in this category apart from another?
- Are sales of these types of products increasing, declining, or staying flat?

- What is the average mark-up for this type of product at your organization?

- Can you tell me whom in your organization makes purchasing decisions for these products?

- Can I say you directed me to them?

- Can you tell me whom your company orders these products from?

- I would like to contact the sales rep at the firm you buy from; can you give me their contact information? May I use your name when I contact them?

Thank you for your time and for sharing your expertise.

Appendix J: Sample Cover Letter to Catalogs:

Dear Prospective Buyer:

I am a product developer selling a new modem device called The Connector, which adds reliable, high-speed Internet access to your existing home or office phone line. This device ends the hassles of dial-up modems and busy signals. Installation is easy: simply plug the Connector into the 10/100 Base-T or USB port of your PC, apply power, perform the simple software configuration, and connect your ADSL phone line to the device.

This modem is capable of data rates hundreds of times faster than a traditional analog modem. But unlike analog modems, the Connector allows you to use the same phone line for simultaneous voice/fax communications and high-speed Internet access, eliminating the need for dedicated phone lines for voice and data needs. No user configuration is required.

I am writing to request you consider including the Connector in your catalog. Enclosed within this envelope, please find a sell sheet, price list, and information regarding the terms and conditions. If you would like to test-market the Connector, we are prepared to offer you the most competitive price from our price list: [$x] per unit, for any size test run you would like. Please send any product submission forms or guidelines. You can contact me directly at [xxx-xxx-xxxx]. Thank you for your time and consideration.

Sincerely,

[Inventor name, company position]

Enclosures.

Appendix K: Glossary of Terms

angel investor: Affluent individuals who capitalize startups in exchange for convertible debt or ownership equity

arbitration: A legal, alternative dispute resolution that settles disputes outside the courts by referring the dispute to one or more impartial legal experts

claims: Part of a patent application that defines the scope of protection granted by the patent

contingent fee intermediary: Experts trained to find suitable manufacturers and distributors for an invention

continuation application: A request to file a patent under new claims, particularly if there is reason to believe the application was not filed accurately

contract manufacturers: Manufacturers that receive outsourced manufacturing jobs from manufacturing licensees

commercialization: The procedure that starts with the genesis of a new product idea, results in mass production, and ends with the customer buying the product

declaratory relief: A request to determine the validity of a patent and determine whether the patent has actually been infringed upon

distribution tier: The path a product travels in order to go from being manufactured to being available on a store shelf

doctrine of equivalents: A legal rule that establishes infringement based upon the similarity of function, method, or results

end user: The customer who purchases the product at the very end of the distribution tier (i.e. consumer who purchases the product in retail stores)

enhanced damages: Deliberate infringement in which the assessment of royalties owned may be tripled

equivalents: Elements of a device or steps of a process that were either available when the patent was issued or available after the patent was issued but before the patent was infringed upon

equivalent infringement: Infringement by which someone designs a process or device around a patent claim in such a way that it functions in substantially the same manner and provides the same result as the patent claim

exclusive license agreement: An agreement by which the licensee acquires the sole right to exploit your patent claims, and no one else can acquire them

fit-and-finish prototype: A prototype that represents a final version and requires no redesign

Gorham test: A test that determines whether an ordinary consumer finds the patented and the infringing design to be similar such that their resemblance is considered to be deceptive to the consumer and induces the consumer to make purchase of the object containing the infringing design

gross profit margin: The difference between the cost of making a product and its selling price

hybrid license: A license that assigns rights from a patent that is issued or pending, as well as an addition form of intellectual property, such as a trade secret or how-to

inducing infringement: A type of infringement in which a party is persuaded to make, use, or sell a patented invention without authorization

industrial designer: A professional who works with inventors to design and develop consumer product inventions

injunctive relief: A court order to cease the infringement and warn of monetary penalties should infringement continue

intellectual property: An original and intellectually invented composition, device, or process that has potential commercial value

key decision maker: Executives in the distribution tier who have the power to make the decision to license or receive assignment to intellectual property rights

know-how: The inventor's expertise regarding the invention that facilitates implementation of the invention

licensing agent: A professional who provides contracted marketing services to get an invention in front of prospective licensees in return for a percentage of the total gross licensing deal

line extensions: New products that complement existing products

literal infringement: Infringement that directly makes, uses, or sells a process or device that contains every element of a patent claim

market value: The price at which a seller is willing to sell a product and a buyer is willing to buy it

Markman hearing: A hearing held in which the patent owner and an infringing party are allowed to present their interpretation of claims to a judge

means-plus-function: A patent claim that specially describes a function of an element rather than the structure of the element

mediation: A form of alternative dispute resolution between two or more parties in which a third party, the mediator, helps to negotiate an out-of-court settlement

net sales: All sales minus returns or nonpayments

nondisclosure agreement: A confidentiality agreement between parties who agree to restrict the access and use of shared information, knowledge, or material from additional parties

nonexclusive license agreement: A license agreement in which the licensor retains the right to license the patent claims to more than one licensee

option fee: Fees paid to the licensor by the licensee for the right to "rent" a product for a certain period time as compensation for possible lost business opportunities during that period

original equipment manufacturer: Firm that manufactures products or components purchased by a company and retailed under that purchasing company's brand name

patent: Exclusive rights granted by a state to an inventor which prevent others from making, selling, or using a product based on the claims made on the invention

patentability: The degree to which an invention's claims meet the relevant legal conditions to be granted a patent

patent agent: A legal professional who has passed the bar exam, is registered with the USPTO, and can be hired to perform prior art searches

perceived value: The manufacturer's estimated cost to acquire a product, plus the estimated minimum profits and retail price that the market will bear

prima facie proof: Proof given of something's existence on first appearance

prior art search: A preliminary patent search conducted to determine the existence of similar claims

proprietary submission statement: An agreement that stipulates a company will agree to review an invention, keep the invention and all attached documents in confidence, to return all documents submitted, and to pay a reasonable sum and royalty if the invention is adopted

prototype: An early sample or model built to test a concept, device, or process

pull-through sales: The amount of sales an individual company estimates a product must have in order for it to be considered marketable

reduction to practice: The practice of making a working model of the invention that demonstrates that the invention will work to fulfill its intended purpose

retail market value: The total cost of a product's materials, manufacturing, packaging, distribution, and marketing

royalties: Payments made by one party to another for ongoing use of intellectual property

specifications: Explanation of what the invention is made of and how it works

stylized format: The specific visual design or scripting of a trademark

test marketing: The method of preliminary testing in the marketplace by which a prototype's sales potential is measured

tiered risk: The amount of risk encountered at various stages of product development

tooling: The process of providing a factory with the equipment needed for manufacturing

trade secret: A composition, process, or device developed by someone and not publicly known

typed format: A trademark that is a word

venture capitalist: An institutionalize version of an angel investor

virtual prototype: A three-dimensional, computer-generated product model

Bibliography

About.com. "Patent Claims." 2011.
http://inventors.about.com/od/patentsbasics/a/PatentClaims.htm.

Answers.com. "Different Types of Retailers." 2011.
http://wiki.answers.com/Q/What_are_the_different_types_of_
retailers.

Bpmlegal.com. "Marketing companies." 2010.
www.bpmlegal.com/pinvmktg.html.

Charmasson, H. *Patents, Copyrights and Trademarks for Dummies*. Hoboken,
New Jersey: Wiley Publishing, 2004.

Docie, R. *The Inventor's Bible*. Berkeley: Ten Speed Press, 2004.

"Evaluating Supply Chain Management." 2011.
www.tompkinsinc.com/publications/competitive_edge/articles/
06-03-Good_Better_Best.asp.

Freeadvice.com. "Trade Secret Law." 2011.
http://law.freeadvice.com/intellectual_property/trade_secrets/
state_trade_csecret_laws.htm.

Ftc.gov. "Scammers who revive their company." 2011.
www.ftc.gov/opa/2007/09/inventionswindle.shtm.

Going-global.com. "Analyzing Foreign Markets." 2005.
www.going-global.com/articles/analyzing_foreign_markets.htm.

Helium.com. "Liability insurance." 2011.
**www.helium.com/items/90440-tips-for-inventors-how-to
-purchase-liability-insurance**.

How To Protect and Benefit From Your Ideas. Arlington, Virginia: American
Intellectual Property Law Association, 1981.

InventionCity.com. "Inventing 101."
www.inventioncity.com/inventing101.html.

Inventionstatistics.com. "Number of Patents Filed Annually." 2011.
**www.inventionstatistics.com/Number_of_Patent_Applications_
Filed_Annually_Year.html**.

Inventnet.com. "Marketing Scams." 2011.
www.inventnet.com/scam.html.

Jpbsonline.org. "Commercialization strategy for Extending Drug Life Cycle." 2010.
**www.jpbsonline.org/article.asp?issn=0975-7406;year=2010;volume
=2;issue=1;spage=2;epage=7;aulast=Gupta**.

Lander, J. *All I Need Is Money: How to Finance Your Invention*. Berkeley:
NOLO, 2005.

Lawfirms.com. "Marketing scam statistics." 2011.
**www.lawfirms.com/resources/intellectual-property/licensing/
licensing-agents.htm**.

Law.upenn.edu. "Uniform Trade Secrets Act." 1985.
www.law.upenn.edu/bll/archives/ulc/fnact99/1980s/utsa85.htm.

Leahy.senate.gov. "America Invents Act." 2011.
**http://leahy.senate.gov/imo/media/doc/PRESS-FirstInventorTo
FileSupport-OnePager.pdf**.

Levy, R. *The Complete Idiot's Guide to Cashing In on Your Inventions*. New
York: Alpha Books/Penguin Press, 2002.

Lventre.com. "Foreign Filing License." 2005.
www.lventre.com/foreign.html.

Manta.com. "Number of U.S. Manufacturers." 2011.
www.manta.com/mb_33_E7_000/manufacturing.

Mcgraw-hill.com. "Product Design: Chapter 4." 2007.
**http://highered.mcgraw-hill.com/sites/dl/free/0072983906/
234967/chase_ch04.pdf**.

Mhlnews.com. "Supply Chain Performance." 2011.
http://mhlnews.com/news/mhm_industrynews_1038.

Netmba.com. "Pricing strategies,"
http://www.netmba.com/marketing/pricing. 2010.

Overholt, S. *New Product Commercialization*. Pennsylvania: Gannon
University Small Business Development Center, 2004.

Patent-trademark-law.com. "Inventor Scam Prevention." 2011.
**www.patent-trademark-law.com/patents/inventor-scam
-prevention**.

PharmExec.com. "Prozac losing market share after patent expiration." 2004. **http://pharmexec.findpharma.com/pharmexec/article/article Detail.jsp?id=136706**.

Pressman, D. *Patent It Yourself.* Berkeley: NOLO, 1985.

Reese, H. *How to License Your Million Dollar Idea: Everything You Need to Know to Turn a Simple Idea into a Million Dollar Payday.* New York: Wiley & Sons, 2002.

Rqriley.com. "Wright Brothers Conquest of the Air." 1999. **www.rqriley.com/wrights.htm**.

Smallbusinss.chron.com. "Comparing Company Business Models." 2011. **http://smallbusiness.chron.com/evaluate-business-model-502.html**.

Stim, R. *Profit From Your Idea: How To Make Smart Licensing Deals.* Berkeley: NOLO, 1998.

U.S. Department of Commerce. *Patents and How to Get One: A Practical Handbook.* Minneola, New York: Dover Publications, 2000.

USPTO.gov. 2011. "Patent Search Facility Address." **www.uspto.gov/products/library/search/index.jsp**.

Virtualpet.com. "Invention Commercialization Process: Marketing an Invention." **www.virtualpet.com/invention/inventionprocess.htm**.

Younkle, M. *Five Steps for Profiling the Market for Your Invention.* Mathew Younkle, 2002.

Author Biography

Michael J. Cavallaro was born in New Hyde Park, New York, and was educated at Villanova University. Following his years as an editor with HarperCollins Publishers, he has worked as a freelance writer for commercial business. This is his third book.

Index